PROBABILITY IN ECONOMICS

Notions of probability and uncertainty have been increasingly prominant in modern economics. This book considers the philosophical and practical difficulties inherent in integrating these concepts into realistic economic situations. It outlines and evaluates the major developments, indicating where further work is needed.

The book addresses:

- probability, utility and rationality within current economic thought and practice;
- concepts of ignorance and indeterminacy;
- experimental economics;
- econometrics, with particular reference to inference and estimation.

Probability in Economics provides a comprehensive overview of one of the most important issues in economics without relying on a mathematical or overly technical presentation. Its conclusion is that views of probability and uncertainty within contemporary economics are much too narrow, and that this severely limits their application to real circumstances and to normal patterns of choice.

Omar F. Hamouda is Associate Professor of Economics at York University, Ontario. He is the author of four books (and numerous articles) and the editor of six others. His interests include macro and monetary economics, the history of economic thought and economic methodology. **Robin Rowley** is Professor of Economics at McGill University, Montreal. He has published many articles in refereed economics journals, has written or edited eight books, and been an active consultant. Trained as an econometrician, his recent interests also include fiscal policies, industrial structures, globalization and educational issues.

ROUTLEDGE FRONTIERS OF POLITICAL ECONOMY

PROBABILITY IN ECONOMICS

Omar F. Hamouda and Robin Rowley

London and New York

First published 1996
by Routledge
11 New Fetter Lane, London EC4P 4EE
Simultaneously published in the USA and Canada
by Routledge
29 West 35th Street, New York, NY 10001

Routledge is an International Thomson Publishing company I(T)P

Typeset in Garamond by Routledge
Printed and bound in Great Britain by
TJ Press (Padstow) Ltd, Padstow, Cornwall

British Library Cataloguing in Publication Data
A catalogue record for this book is available from the British Library

Library of Congress Cataloguing in Publication Data
Hamouda, O. F.
Probability in economics/O. F. Hamouda and Robin Rowley.
p. cm. – (Routledge frontiers of political economy, ISSN 1359–7914)
Includes bibliographical references and index.
1. Economics, Mathematical. 2. Probabilities. 3. Uncertainty. 4. Utility
theory. I. Rowley, J. C. R. II. Title. III. Series.
HB135.H354 1996   95-46826
330' .01 '5192–dc20   CIP

ISBN 0–415–06712–X

CONTENTS

INTRODUCTION

Economists have increasingly used the language and symbols of probability through the last three decades. The earlier reliance on multiple regression and correlation – with a small menu of methods for estimation of structural components in probability models, for drawing statistical inferences and making numerical or qualitative predictions – has widened to deal with a host of additional complications. New interests include the recognition of limited ranges for some economic variables, awareness of their latency and incomplete measurement, the cointegration of economic variables over time and concern over difficulties stemming from the inherent structural instability of socioeconomic phenomena. Besides this widening and rapid growth of complexity, empirical research seems to have been overwhelmed by the flood of statistical testing (to the extent that we are accused of 'star-gazing' by sceptics who doubt both the foundations of these tests and their awkward interpretation by economists).

Contemporaneously, with the growth of technical training and perhaps with excessive zeal, some prominent theorists have moved beyond the simple linear difference equations of earlier years by adding stochastic errors (reflecting supply shocks, changes in technology, monetary surprises and other exogenous influences). These influential economists have effectively pushed a few time-series notions into professional consciousness. Macroeconomics, for example, has been transformed by a new radical reliance on 'rational' expectations, ergodicity, stationarity and potential 'unit roots'. In a parallel development, other theorists have expanded their models to include simple distributional elements. Such elements now represent important aspects of risk aversion and stochastic dominance in theories of choice and are used to describe the processes assumed to govern basic search behaviour of individuals in labour markets.

1

Meanwhile, the validity of some aspects of rational microeconomic theory (such as transitive preferences and responsiveness of competition) were questioned as interest grew in the uncertain context of economic behaviour. Reminiscent of developments during a brief period in the 1930s and 1940s when uncertainty was actively recognized (stimulated by the efforts of Hayek, Shackle and Hart for example), microeconomic 'foundations' are being transformed by an awareness of persistent uncertainty or novelties and the potential consequences.

Given this strong 'emergence' of probability in economics, the present time seems particularly apt for more exploration of the troublesome integration of probability and uncertainty in economics. Clearly, we need to identify and clarify the main lines of important theoretical and empirical developments, to recognize the potential hazards of an excessive reliance on our inadequately-understood probability notions and to explore the idiosyncratic (but interesting) views of prominent economists, such as Shackle and Hicks, who resisted major features of these recent developments or their antecedents. In an earlier book on the history of economic ideas and practices, *Expectations, Equilibrium and Dynamics* (1988), we revealed our strong dislike of the excessive compartmentalization of economics and we adopted a historical perspective. The latter choice was made in the hope of reducing a common forgetfulness of the past among economists, which threatens to divert and constrain the progressive evolution of economics.

These biases are retained here for the current exploration of probability and uncertainty. We freely draw on econometrics, economic theory and decision analysis as if they share common ground – although they are typically separated in university courses, in comprehensive fields of study and in the distinctive focus of most professional journals. Furthermore, we continue to stress the benefits that accrue from looking at current fashions among the rival schools of economic thought with an awareness of their partial dependence on the earlier interests of economists. Finally, as in the other book, we deliberately make very modest demands on mathematical backgrounds of readers – although there is a fundamental need for adequate mathematical skills to fully appreciate the topics considered here. A lack of mathematical preparedness should not prevent anyone from seeking to explore the relevant substance of recent developments in economics and to evaluate the progress being made toward achievement of a better understanding of actual economic phenomena. We hope to promote search by providing a suitable, if provocative, form of access to a wide number of themes with *mathematical requirements reduced as far as possible.*

There are, of course, significant issues raised by Clark (1947) as to whether mathematical economists and econometricians themselves should improve the communicability of their results to a wider audience and give more attention to the tenuous connections between all abstract symbols and the complexities of economic reality. Such issues have been linked by probabilists with an evident need for the specification of *clear domains of applicability* for the axiomatic schemes of probability theory, for regression or causal models and for methods of statistical inference. With respect to the first of these elements, for example, Good (1957: 21) indicates 'there should be precise rules of application of the abstract theory that give meaning to the undefined words and symbols', an obligation rarely honoured in practice. Possibly, we need to keep some empirical standard in mind for our work as economists.

Distinguished probabilists such as Neyman and Jeffreys, both citing Karl Pearson's *Grammar of Science*, felt it worthwhile to stress views that we must connect our scientific models with the world around us, and that the models are frequently inadequate representations of our world. This intuitive and empirical connection is to be reflected both in axioms that we invoke to set the frameworks for formal analyses and in basic choices of language, symbols and mathematical properties through which the confrontation of outcomes of models with the real world is made possible. The empirical validation is acknowledged, for example, when Neyman suggests:

> A model is a set of invented assumptions regarding invented entities such that, if one treats these invented entities as representations of appropriate elements of the phenomena studied, the consequences of the hypotheses constituting the model are expected to agree with observations. If, in all relevant trials, the degree of conformity appears to us satisfactory, then we consider the model an adequate model.
>
> (1957: 8)

In Chapter 1, this view and its counterparts from various schools of probability are clearly present when the diversity of probabilistic notions is identified and when axiomatic schemes for probabilities, conditional probabilities and expectations are noted. Diversity is expressed by reference to taxonomies offered in major commentaries from different perspectives – commentaries that should be consulted for more information. Often, in this context, the discussions by economists are very limited in scope. They can give the false impressions that probability is readily identified with stable relative frequencies of potential economic events and that probabilistic concepts are dichotomized between objective chances (perhaps relative frequencies again) and alternative subjective views, both assumed to satisfy a well-known set of axioms for consistency. We avoid this myopia by widening our perspectives – alternative ingredients are briefly outlined; some notions of rationality, coherence and hierarchy are clarified; and the daunting dynamic and practical problems of elucidation, inaccuracy, convergence of probabilities, emergence of consensus, and similar issues in economic contexts are noted. At times, our survey may resemble an annotated bibliography since *we rely on other references being used to supplement* our terse presentation. Finally, we ask whether, given the severity of such problems, any probabilities are 'needed' or useful – drawing on discussions by Fine (1973: 249-50), Glymour (1983, 1989), Levi (1974) and Shackle. Throughout this chapter, as in all parts of the book, we point to many lines of argument *without offering solutions or advocating the adoption of a particular perspective*. We are not seeking to demonstrate the superiority of one favoured line of argument since this would be premature.

In Chapters 2 and 3, we focus on different aspects of expectations which have enjoyed a prominent place in both theoretical and applied treatments of probability since Huygens discussed their measurement and foundations in the middle of the seventeenth century (Hacking, 1975; Holgate, 1984). Expectations have evolved in economic models from earlier vague anticipations or forecast values (Bigg, 1990) – first to means and modes of probability distributions and then to formal characterizations of random time-series processes. The brief account in Chapter 2 begins with some historical discussions of 'fair games', which provided the immediate stimulus for formal expressions of the mathematics of chance, and with simple notions of risk aversion. The St Petersburg paradox and introduction of 'expected utility' in hypothetical gambles with ill-defined gains are considered in relation to the hazards of using an ephemeral concept (utility) with unclear cardinality. This leads to an account of the issues and paradoxes associated with games after reformulation by von Neumann and Morgenstern in the 1940s and in subsequent revivals, ending with the Allais paradox.

In a parallel historical development, outlined in Chapter 3, another use of expectations became embedded in the choice of assets for wealth portfolios. This shift to aggregates drew attention to a practical need to consider the complementarities of individual risks and thus to recognize the second (and higher) moments of joint probability distributions, defined over the ranges of financial values for the assets rather than individual measures of location or their certainty-equivalents. The shift is traced from its first emergence in the 1930s. We then explore the presence and use of 'rational' expectations at market levels. Initial recognition of such expectations began, surprisingly and belatedly, as a terse postscript to the search for approximate or normative decision rules – a search stimulated by the emergence of operations research or management science.

Economists have used the concept of rational expectations to reconcile objective chance and subjective probability within theoretical frameworks involving 'representative' individuals and implicit disciplinary mechanisms (which reduce or remove any irrational activity), despite the awkwardness of heterogeneous information and differential learning. The chapter ends with a short look at the potential confusion of procedural rationality with its substantive counterpart. The content of this section is heavily influenced by Simon, who made us aware of the need to be more ambitious by drawing on efforts of psychologists to clarify cognition, bounded human rationality and computability. When elucidation of subjective probabilities is attributed to mental and social activities, it should reflect such activities.

In Chapters 4, 5 and 6, we deal with interesting qualifications to probability models that involve unconventional considerations – ignorance, surprise and vagueness – as well as looking at novel perspectives addressing the nature of evidence and support in uncertain situations (perhaps when the full range of probabilities is imprecise) and the dynamics of beliefs when probabilities are revised and their referential frames modified. Here, we return to issues that arise when group choices are associated with a collection of probabilities or preferences for any group as a whole. These issues include the haphazard nature of dialogue, processes of consensus building, and the feasibility of convergence from initial divergence within a group.

Views of Shackle, Zadeh and concerned researchers who have explored the awkwardness of determining valid social criteria for unacceptable risk are the primary focus of Chapter 4. Shackle offers a very effective criticism of frequentist notions applied to unique decisions by entrepreneurs and gives an alternative framework involving potential surprise. Zadeh has stimulated interest in vagueness by promoting notions of fuzziness. We also note the rapidly growing literature on risk in relation to major environmental, medical and socioeconomic problems. Here the connections with some well-established probabilities of disasters strains credulity without removing the urgency for appropriate risk assessments. In recent years, the format of earlier cost–benefit analyses of major public works, industrial innovations, medical programmes and other substantial projects (such as stressing the optimization of expected present values) has been replaced by an effective weakening of precise probability models that reflects practical problems of measurability and evaluation.

Experimental irrationality, as reflected in non-transitive choices and cognitive illusions, contextual factors in decisions (linked to the ambiguity of signals and framing of situations) and some dynamic concerns – found in new theories of prospects, regret and disappointment, and treatments of sequential factors and intertemporal substitution – are discussed in Chapter 5. The emergence of substantial interest in irrationality, context and potential dynamic models of choice and preference serves to undermine some comfortable features of stability, coherence and consistency in simple probabilistic models.

Hayekian views on knowledge and the problem of incomplete probabilities, as raised by Keynes and Hicks, provide the initial focus of Chapter 6. These aspects are widened to include other treatments of evidence and support, including clarification of various approaches to the 'weight of evidence' and the support functions of Shafer and Dempster. For subjective or personalist perspectives, there is also a need to recognize the instability of personal and group-based probabilities, and even the avoidance of numerical assignments.

Almost all training programmes in economics contain a firm requirement for their participants to achieve a modest knowledge of basic estimation techniques, popular software packages and some inferential methods or 'tests'. Econometrics provides the focus of Chapter 7. For convenience, the discussion falls into two distinct parts. The first of these sections deals with four matters: conventional approaches to testing, especially the Neyman–Pearson format and significance testing; the nature of the 'probability approach' to econometrics as developed at the Cowles Commission in the 1940s and 1950s; some aspects of 'non-structural' or time-series modelling; and the apparent neglect of Bayesian approaches in econometrics.

The second section of Chapter 7 deals with three radical reappraisals of econometrics, which challenge normal practices and language of most econometricians over the last half-century. These three initiatives are loosely attributed to three of the principal proponents of fundamental change: Hendry, Leamer and Kalman. The section also considers other novel perspectives, including advocacy of 'without too much theory' (a rejection of excessive *a priori* specification in econometric models), Lawson's views of instrumentalism and a preference for calibration of parametric values in probability models rather than estimation.

In Chapter 8, we provide a brief overview and offer some concluding remarks. Our modest objectives are re-expressed and some major omissions are identified without apology. We remind potential readers that *our modest effort is not intended to be explored in splendid isolation.* Throughout our treatment, we provide information on other sources which should be explored. Chapter 8 contains a selection of our favourite sources for broadly-based information, such as Eatwell *et al.* (1990a), Hacking (1965, 1975) and Keynes (1921). These sources illustrate the rich literature on probability that exists to be explored for both pleasure and deeper awareness.

Our choice of topics in the eight short chapters is directed towards the troublesome aspects of probability used by economists. In part, this choice reflects the existence of a large number of excellent texts describing the main features of probabilistic models, their manipulation and theoretical outcomes. The choice also reflects concern that economists and other social scientists often neglect significant philosophical underpinnings of their efforts to develop models with probabilistic ingredients. It now seems clear that *most economists need to pause to think about the broader aspects of what they do* in this context. Some may find an enlightenment which affects their motivation, the questions they address and the techniques they use. Perhaps, we may see less concern for ephemeral fashionability and more interest in 'scientific progress' if this does not seem a naive objective.

1

VARIETIES OF PROBABILITY

Many economists and other social scientists adopt a very casual approach to the meaning and origins of probabilities. They almost always seem quite content to preserve a comfortable familiarity with some basic probabilistic concepts – such as distribution or density functions with convenient mathematical properties, reasonably stable parameters, simple stochastic processes with sufficient stationarity over time and a few referential statistics for both location and dispersion (e.g. mean, median or mode, and variance or standard error). Probabilities themselves are generally separated only into 'subjective' and 'objective' types without much consideration of what features these qualifiers should entail, other than some notion that one type might just involve individual or personal mental activity while the other type reflects an external or contextual property. Both types are presumed to obey a group of similar (perhaps identical) rules for consistency, completeness and coherence.

Economic theorists, in particular, persistently assume the existence of numerical or hypothetical probabilities while they fail to address the issues associated with their measurability, the feasibility of consensual convergence on particular values within groups, any potential spatial or historical instability and the possibility of elucidation in a wider sense – although terms like information, knowledge and learning are common enough. When found convenient for narratives or theoretical analyses, the objective and subjective probabilities are equated or confused and the subjective probabilities (surely personal in their origins) are often attached without any limiting clarification to groups, even to complete markets, in models that otherwise are associated with rigour. Econometricians and sociometrists generally waste little time on explaining, rather than simply using, the familiar inferential statistics and standard methods of estimation with their unfamiliar bases for validation or interpretation – as, for example, in the common confusion of statistical significance with substantive significance, the use of asymptotic justifications for parametric estimates in the real context of socioeconomic phenomena that are clearly unstable in the long run, and the ignorance of pretest biases and other consequences of sequential modelling beyond a simple criticism of apparent 'data mining'.

This 'untidy' picture of neglect, vagueness and technical myopia stems in part from a firm perception that the past and current features of behaviour and priorities have been relatively successful, even progressive, and thus little change is needed. Some mathematicians, in weaker fashion, argue that the 'theories of probability based on different philosophical concepts may lead in practice to the same or almost the same results' (Renyi, 1970: 34). However, beyond such hopeful assertions of immanent standards, it also shows an inadequate reflection on the fundamentals of probabilities, including their variety, which affect their meaning, validity and extension to any meaningful areas of inference, decision and prediction.

Clearly, very little harm is done by acknowledging this inadequacy and briefly exploring some of the important aspects of probability that have often been neglected. In the following four sections: various types of probabilities are identified by reference to some of the important taxonomies that have been put forward; a few axiomatic schemes or basic rules for mathematical operations are linked to alternative elements of probabilities, conditional probabilities and expectations; an appreciation of major areas of concern (rationality, coherence and elucidation) is raised; and basic issues are linked to hierarchies of uncertainty, overspecification and interpretation.

VARIOUS NOTIONS

The use of the term 'probability' and its counterparts has become commonplace in much scientific literature. Within social sciences, including economics and the humanities, the language of probability has moved far beyond any simple identification with a ratio of two numbers in connection with 'odds' in games of chance, and its gradual assimilation in wider areas has continued unabated throughout the last half century. However, all serious attempts to step beyond the familiar acceptance to some deeper requirement of precise meanings or unique definitions for our probabilistic notions have inevitably led to severe practical and philosophical problems. Given a recognition of

9

such problems, it is hardly surprising to discover the existence of alternative taxonomies for the various notions of probability – e.g. as proposed or described by Bunge (1988), Carnap (1950), Cohen (1989), Fine (1973), Good (1965, 1983), Hartigan (1983), Nagel (1939), Savage (1954, 1977) and Weatherford (1982).

Should the apparent non-convergence of views on probability, as illustrated in these taxonomies, matter to economists? Many responses to this question stress the linkage between potential areas of *employment* of probabilistic notions and *justification* of the various concepts. For example, Cohen (1989: 41–2) explains the basis for his treatment:

> we explain and justify the existence of a variety of legitimate types of probability judgement by exploring the wide range of syntactic and semantic factors that are relevant to the taxonomy and employment of such judgements. Only when this task has been adequately carried out will we be in a proper position to discuss what types of probability-judgements are suitable – and under what conditions – for employment as a mode of inductive evaluation.

There would be no need for probability if there were 'ready-made' and unambiguous answers (such as right or wrong, true or false, and certain or impossible) whenever an assertion is made about the world around us or a statement is made about some abstract hypothesis. However, assertions of these kinds are generally made when the comprehensive knowledge required for assessment is incomplete, insufficient and unavailable. Some ambiguity then characterizes the immediate environment for assessment and our various notions of probability may provide tentative bases for '*in-betweenness*' of the binary alternatives (right–wrong, true–false, certain–impossible). We opt for elusive notions of 'more or less probable'. Concepts of probability are then convenient abstract devices with which to indicate some degree of confidence in a statement or to 'measure' the confirmation of a factual hypothesis in a situation of incomplete or unavailable knowledge.

Whether a probability is attached to factual events or to statements of abstract hypotheses, effective evaluation requires either the investigation of similar instances or some deduction with mental arguments. Probabilities can thus be empirical or logical, inductive or deductive. Typically, any meaningful evaluation draws on evidence, intuition, conviction and other ingredients for its support, so the resulting probabilities can be objective or subjective. The search for relevant support, even when dealing with a unique case, clearly involves some reference to similar or comparable instances and analogies. Whenever there are sufficient prior instances, the ideas of past recurrence or future replication provide a sensible basis for using the notion of relative frequencies for the evaluative process. Otherwise, judgement may depend on the elucidation of beliefs. Probabilities may then be defined by reference to the frequencies or to degrees of belief. Finally, probabilistic evaluations will differ with respect to their reliance on the degree of measurability and the need for perceived and practical cardinality or ordinality.

The varieties of probability may be empirical or logical, inductive or deductive, objective or subjective, cardinal or ordinal, precise or boundaries, individual or multi-subjective, and frequentist or belief-driven. Thus, it is not surprising that many alternative definitions of probability have emerged with a host of different qualifiers or labels – e.g. physical, material, intrinsic, psychological, logical, intuitive, tautological and epistemological. Sorting through the bases for such labels must inevitably lead to substantial differences among the simplifying taxonomies that have been put forward.

Beyond simple labels, the clarification of our probabilistic notions should recognize the common presence of both realistic and idealistic elements:

> even well-argued analytical theories about the nature of probability are not always targeted at exactly the same objective. One polar aim. . .is to describe, analyse and explain how probability is actually conceived in human judgements. The alternative polar aim is to prescribe how probability should be conceived and. . .how the formalism of the mathematical calculus should be interpreted.
>
> (Cohen, 1989: 41)

The distinction between these two polar aims arises whenever we seek to clarify the various reasons for estimating or asserting probabilities – bets, decisions, discrimination, predictions, insurance, confirmation, determining surprises, improving understanding (Good, 1965: 3).

The prescriptive aim became more prominent in economics some forty years ago when Savage, Friedman and others (influenced, in part, by the rekindling of interest in games, bargaining theory and rational conduct due to von Neumann and Morgenstern) sought to describe risk aversion and to extend subjectivist approaches to probability. A confusion between the polar aims seems to have proliferated in the recent use of rational expectations, where the descriptive and pre-scriptive languages of probability are often used interchangeably! These matters are considered in the next two chapters.

The classifications in taxonomies laying out the alternative con-cepts of probability have grown over time. Nagel and Carnap, writing in the late 1930s and early 1950s, could manage their accounts with just two or three classes beyond the classical or Laplacian concept. Since then, plurality has been generally accepted and more classes explored – with four classes favoured by Good, Schoemaker and Weatherford, more by Cohen, and Fine (1973) who opted for no less than eleven classes.

Carnap (1950) insisted that all inductive reasoning involves a 'prescientific concept' of probability. He separated the logical seman-tic notion of probability$_1$, associated with the degree of confirmation of any hypothesis with respect to some given evidence or premises, from the factual or empirical notion of probability$_2$, the long-run frequency of a property of events or things. The latter concept, while unsuitable for inductive logic, was held to be useful in other respects. Carnap dismissed controversies among the different schools of pro-babilists as futile, erroneous and unnecessary if they fail to recognize that various notions of probability may be intended to serve different purposes and that no notion is suitable for all purposes. Nagel also distinguished probability as a unique logical and objective relation between propositions – for example, the constructive theories of Keynes and Jeffreys – from the measures of relative frequencies as developed by von Mises and Reichenbach.

For present convenience, with a small degree of distortion (avoided by exploring the greater details given in the references cited), it is useful to focus attention on four basic strands in popular taxonomies – identified by common qualifiers such as classical, frequentist, logical or necessarian and subjective or personalist – rather than go into any finer detail. With this simplification, sufficient varieties of probabil-ity still emerge very clearly. This simplifying approach is not novel with us but, as acknowledged by Savage (1954: 3) who followed a similar path, it 'is bound to infuriate any expert on the foundations of probability'. Fortunately, the expert was never intended to be part of our audience!

The Laplacian or classical theory of probability, sometimes called the indifference theory (Cohen, 1989: Fine, 1973; Keynes, 1921; Nagel, 1939; Weatherford, 1982), prevailed for almost two centuries. It required the elementary potential outcomes to be 'equi-probable' (an inevitable source of circularity if viewed as realist rather than idealist) and it associated probability with the numerical ratio of favourable outcomes to the total number of potential outcomes, where the term 'favourable' refers to the occurrence of some specific attribute of interest. This classical conception seems to have originated with Jacob Bernoulli at the beginning of the eighteenth century, partly from 'the (ideal) physical symmetry of decision instruments common to games of chance' (Bynum *et al.* 1981: 338). Its rapid assimilation was enhanced by de Moivre, and it was subsequently developed by Laplace and his followers at the beginning of the nineteenth century. Although effectively constrained by the equi-probable assumption, the concept was convenient when linked with the mathematics of permutations and combinations, and it always gave numerical and unique values. When any probabilities were unknown, Laplace provided a simple objective mechanism for the elucidation of these values – provided the elemental equi-probable format existed and was identified. Unfortunately, the attractiveness of this format stimulated some vigorous excesses with the supplemental presumption that ignorance may reasonably imply equally-likely elements ('rectangular or uniform priors' in modern statistical language) – an instrumental presumption which permitted paradoxes or inconsistencies and introduced implicit subjectivity. An explanation for the persistence of the equi-possibility format and an informative account of its support and criticism are provided by Hacking (1975).

Although both Mill and Jevons questioned the interpersonal stability of any numerical probabilities based on judgement and the classical format, it was left to Keynes (among economists) to launch a major assault on the concept. He recognized (Keynes, 1921: 417) that the inductive Laplacian process, creating some posterior probabilities from elemental equi-probable ingredients, had already been 'rejected by Boole on the grounds that the hypotheses on which it is based are arbitrary, by Venn on the ground that it does not accord with experience, by Bertrand because it is ridiculous, and doubtless by others also'.

By the middle of the nineteenth century, new attitudes to probability had begun to emerge, involving observation, group phenomena, and experience. Ellis, for example, argued that 'mere ignorance is not grounds for any inference whatsoever' while Venn insisted that 'experience is our only sole guide' (Keynes, 1921: 93). The gradual spread of these new attitudes and the emergence of statistical analyses undermined acceptance of the classical concept, and they prepared the ground for both the popularization of frequency concepts and the apparent mechanization of knowledge (Gigerenzer, 1987) in the present century. Keynes, however, opted for a logical construction of probability – one which is largely discounted today, and was perhaps eventually discarded by Keynes himself – and he cast serious, if belated, doubt on the ubiquitous measurability of probability.

Logical or necessary and semantic theories of probability (as developed in many different ways by Keynes, Jeffreys (1948), Koopman (1940a), Carnap, and others) generally deal with the existence and use of objective measures of rational support for propositions:

> These authors argued that a given set of evidence bears a logical, objective relationship to the truth of some hypotheses, even when the evidence is inconclusive. Probability measures the strength of this connection as assessed by a rational person.
>
> (Schoemaker, 1982)

> Necessary views hold that probability measures the extent to which one set of propositions, out of logical necessity and apart from human opinion, confirms the truth of another.
>
> (Savage, 1954: 3)

Logical probabilities, sometimes also called credibilities or legitimate intensities of conviction (Cohen, 1989; Fine, 1973; Good, 1965; Levi, 1979; Savage, 1977; Weatherford, 1982), seem unlikely to be known in practice even if they are unique in principle. Most of their proponents invoke, implicitly or otherwise, assumptions of individual rationality in the sense that necessary connections would be recognized and acted upon. Knowledge, often relative to individual experience, is regarded as fundamental and rational belief is defined by reference to it (Keynes, 1921: 10).

The normal scope for objective logical probabilities is unclear – whether we consider only economic phenomena or cast our net somewhat wider. Keynes may have believed that they could apply to many situations but he severely restricted other aspects of them. For example, he conceded that 'the perceptions of some relations of probability may be outside the powers of some or all of us' (Keynes, 1921: 19), that there are many cases in which the probabilities may not be comparable, and that a numerical measure of degrees of probability 'is only occasionally possible' (ibid., 122).

Similar complexities adversely affect the other views of logical probabilities, although some views may be more sanguine about measurability. For example, Koopman showed that it is quite possible to numerically represent some groups of partially-ordered (logical or intuitive) probabilities. Fine provides a summary of axioms for the comparative probabilities of Koopman, and he clarifies the potential connections between them and relative frequencies, while Good (1962a: 322) uses Koopman's notion of upper and lower bounds for probabilities to explain his own attempts to deal with non-measurability.

Jeffreys, who like Keynes was influenced by the Cambridge philosopher W.E. Johnson, and Carnap both accepted numerical probabilities. Prominent advocates of subjective or personalist concepts, like Savage (1977: 13) and Zellner, often recognize formal similarities between their own approaches and those of Jeffreys – he indicated standard prior distributions and some useful rules for constructing probabilities (Hartigan, 1983: 3). Carnap's theories, which suffered from inconsistencies, sought gradation for inductive inference without unique systems of measurement but were compatible with some semi-formal axioms for (quantitative) logical probability or degree of confirmation (Fine, 1973: 187).

The passage of time has generally been quite unkind to most logical or necessary theories. A partisan (but fair) postscript to the earlier interest in the theories is given by Savage more than half a century after the new perspective of Keynes was launched.

> Necessarians are usually temperate in their claims, acknowledging their theorems to be seriously incomplete. Non-necessarians may not find even the ostensible beginnings of necessarian theories to be cogent.
>
> (Savage, 1977: 9)

In similar vein, Bunge (1988: 33) points out that the logical theory 'has found no applications in science or in technology' and its interpretation has been 'absorbed by the more popular subjectivistic (or personal) interpretation'.

The emergence of frequentist views of probability (Kruger *et al.*, 1987a) effectively began in the middle of the nineteenth century through the efforts of Ellis and Venn (Keynes, 1921: 100). Their assimilation was encouraged, and hindered too, in the present century by the persistent attempts of von Mises and Reichenbach to resolve some awkward features of relative frequencies. Acceptance was also helped by statistical theorems which linked tendencies of sample statistics with basic population parameters of probability distributions underlying the samples – theories for 'laws of large numbers' that became part of most introductory courses in statistical education (Bynum *et al.* 1981: 339). From the initial theorem of Jacob Bernoulli, dealing with the deviations of successive sample means from their population counterparts (Hacking, 1975), to the recent explorations of ergodicity and stationarity for long-run sequences, the pursuit of mathematical probability appears to have actively pushed the popularity of the frequentist bases for probability.

The stimulus of Ellis and Venn came from their statistical arguments that probabilities are concerned with series or with groups of events and that their numerical determination must be empirical – a sharp break from the 'a priorism' of the Laplacian formulations, but it came with the additional baggage of major presumptions (including both the independence and randomness of observations, long-run stability, potential replication and the existence of particular population parameters). Clearly, the simple gaming analogy was extended to the statistical outcomes of conceptual experiments and stochastic processes (Feller, 1968), as clarified when von Mises made explicit the notion of a sample space, separate from the basic population, and when Kolmogorov put this space at the heart of his axioms for probability.

For some prominent participants, the scope of statistics (and especially of statistical inference) became concerned *only* with groups and sequences, governed by laws of large numbers and ergodicity, and not at all with single individuals or events (Bartlett, 1975). Others, accepting the frequency interpretation and extending the mathematical theory as relevant to idealized relative frequencies, insisted that such theory must be liable to empirical verification with observable frequencies for it to have practical value (Cramer, 1946; Feller, 1968; Neyman, 1957).

These limitations or qualifications reflect the efforts of von Mises (1957) to establish a sounder basis for the frequentist concept. Beginning in 1918, he kept the earlier focus on both repetitive events and mass phenomena, insisting:

> The rational concept of probability, which is the only basis of probability calculus, applies only to problems in which either the same event repeats itself again and again, or a great number of uniform elements are involved at the same time. . . . [We] may say that in order to apply the theory of probability we must have a practically unlimited sequence of uniform observations.
>
> (von Mises, 1957: 11)

16

To this assertion, he added a condition of randomness, the need for a clear context as specified by distinct 'collectives', and the new view that all probabilities are only the *limiting* values of some observed relative frequencies within such collectives as observations are continued indefinitely. Any suitable collective must meet two strong conditions – the existence of the limiting value and its independence across arbitrary partial sequences (von Mises, 1957: 25) – which are corroborated by experience with games of chance, but need also to be corroborated in different non-game situations.

For the natural sciences and some insurance or demographic problems, the strict requirements of von Mises might be met in practice (at least approximately), but their potential fulfilment in economics is doubtful. Knight (1921), in a very influential book, concluded that they would indeed not be satisfied for most entrepreneurial decisions and profits, leaving room only for anarchic uncertainty which could not be identified with either actual relative frequencies or their limiting counterparts, and eventually. leaving room for normative models with subjectivist ingredients (Arrow, 1951; Fellner, 1965).

Major problems persist even when more optimistic views of practical certainty prevail for the long-run convergence of some relative frequencies. Laws of large numbers do not mean that, at any given moment, the current sample average is equivalent to the mathematical expectation or mean (Feller, 1968: 152), and the calculation of a finite sample average in a given situation does not mean that the suitable limit exists for probability measure here. Similarly, arbitrariness often pervades the presumption of randomness, the quality of predictions, and the choice of many reference classes. Nevertheless, the frequentist concepts of probability clearly influence much of the econometric literature, especially in relation to the properties of estimated parameters and to familiar tests of statistical significance (considered in Chapter 7 below), and many idealistic or normative economic models. The standard elements of various frequentist views (due to von Mises, Reichenbach and Salmon, among others), the major criticisms of them, and some valuable assessments of both their history and practical relevance are described by Bunge (1988), Fine (1973), Hacking (1965), Hartigan (1983) and Weatherford (1982), while the primary treatment of von Mises (1957) remains fascinating.

Soon after Keynes completed his broad history and treatment of logical probability, Ramsey (1926) introduced a fundamentally new approach which combined preference and probability in a clear subjective format. Unfortunately, this approach received little attention until the similar views of de Finetti and Savage (and others) began a fundamental transformation of the acceptability of subjective, personalist or Bayesian approaches to probability, launched a revival of the Bayes's theorem as the basic focus of hypothesis revision, and also re-opened the search for a wider framework of applications to uncertain situations. As described by de Finetti (1972), Pratt *et al.* (1964), Savage (1954, 1961), and Schlaifer (1959), the new perspectives produced a strong momentum from which non-frequentist views became very influential among statisticians, and could even threaten the primacy of frequentist views (at least in some theoretical exchanges). The accounts of Cohen (1989), Fine (1973), Good (1965), Hacking (1965), Schoemaker (1982), Weatherford (1982), and contributions to Eatwell *et al.* (1990a) again separate the subjective conceptions from alternative ones.

Personalists or Bayesians took probability to be 'a certain measure of the opinions of an ideally coherent person' (Savage, 1977: 10). Although their theories could be interpreted from both psychological and empirical positions, observations from psychological research weakened the empirical foundations and a normative emphasis often became predominant – theories describing, from this interpretation, what individuals should do. Given the direct reliance on opinion or belief, interpersonal differences are possible but, 'in practice, it is extremely valuable to note situations in which a wide variety of users will have common opinions, or at least common features to their opinions' (ibid.: 14). The existence of these similarities will then widen the scope for generalization of inference – with probability, axioms of consistency, and hypotheses providing the language and methods of numerical calculation, and clarifying the nature of admissible decisions.

There are many forms of subjectivist theories (Good, 1983) but all seem to accept the meaningfulness of talking about the probability of a hypothesis (the degree of belief in it) and they usually stress attempts to make judgements consistent in some sense – often referring here to the avoidance of a 'Dutch book' in which the odds mean that someone must inevitably lose (Good, 1965: 6), as described by Ramsey. The systematic updating of such probabilities from prior values to posterior ones is then handled through the Bayes theorem. A common view is that this process has wide applicability, much wider than the areas which accept a frequentist format:

18

Probability as degree of belief is surely known by anyone: it is that feeling which makes him more or less confident or dubious or sceptical about the truth of an assertion, the success of an enterprise, the occurrence of a specific event whatsoever, and that guides him, consciously or not, in all his actions and decisions.

(de Finetti, 1974a: 122)

This view should not obscure the realistic objections to degrees of belief. Menges (1973: 8) summarizes five of these objections. Interpersonal communication of such beliefs is limited. Evidence from experiments shows 'the strength of dispositional belief, or introspective opinion, or willingness to bet' may systematically differ from objective probabilities. Contextual influences, such as emotions and intellectual circumstances, have uncontrollable effects. The separation of 'close' probabilities is difficult in practice. Measurement of belief is complicated so an untrained individual might not readily obtain values. Clearly, a shift to idealist or normative interpretations may weaken the impact of such empiricistic objections. Attention can also be redirected toward various means of facilitating the emergence of coherent probability assessments, to the provision of better methods for communication, to clarifying efficient processes of combining information and debiasing, and to determine sensible feedback mechanisms (Lindley *et al.* 1979).

The confident advocacy of Bayesianism for statistics, as a unifying framework and an inevitable basis for sound inference, is clearly expressed by de Finetti (1974a) and Lindley (1975). Other aspects are considered both in the next two sections and in Chapter 2, when we briefly deal with economists' treatment of expected utility, risk aversion, the criteria for rationality in games, and stochastic dominance and preferences. The extension of the subjective theories to group decisions are assessed at various points below and the very modest intrusion of Bayesian approaches in econometrics is noted in Chapter 7. Further criticism and expressions of tentative support for the theories are left to these other parts of our account.

AXIOMS

Until the mid-1920s, probability was not generally treated as a mathematical subject (Cramer, 1981; Doob, 1961, 1976) and there was some confusion between the mathematics of probability and 'real-world' probability. A radical breakthrough occurred when

Kolmogorov (1933) launched the measure-theoretic approach to probability theory. This novel approach – building on the efforts of other members of the Russian School (Maistrov, 1974), Borel's use of denumerable probabilities (Knobloch, 1987), and Lebesgue's measure theory – began with a collection of axioms linking elementary and random events with a field of probability in a special application of additive set functions. The form of the axioms clearly reflects an idealization of long-term relative frequencies, and Kolmogorov (1956: 3) linked them to the earlier efforts by von Mises 'in establishing the premises necessary for the applicability of the theory or probability to the world of actual events'. However, he noted the strength (and weakness) of the axiomatic approach by recognizing that abstract axiomatic theory admits 'an unlimited number of concrete interpretations besides those from which it was derived' (ibid.: 1). Nevertheless, it seems that many of the later probabilists who popularized the rapid assimilation of the axiomatic approach – such as Cramer, Doob, Feller, Loeve and Neyman – felt that the corresponding mathematical probabilities 'have their counterparts in observable frequency ratios, and any probability number must, in principle, be liable to empirical verification' (Cramer, 1946: 151).

Kolmogorov's six axioms are quite straightforward. One axiom requires the set of random events to be a field in the sense that sums, products and differences of any two events will belong to the set of events. A second axiom indicates that the collection of elementary events (from which random events appear as subsets of these events) itself belongs to the field. Three axioms deal with the assignment of non-negative real numbers to reflect the probability of any random event: initial assignment of the individual probabilities; a normalization so that the probability of the certain event is unity; and additivity of probabilities of events without common elements. Finally, for any infinitely decreasing sequence of random events, Kolmogorov indicated an 'axiom of continuity' which attaches a limiting probability of zero to the product. Here, in more technical language, the probability space has three fundamental components: a sample space of random events, a sigma-field of subsets of the sample space, and a probability measure, with the latter governed by some operational rules of normalization, non-negativity, finite additivity and continuity (Fine, 1973; de Finetti, 1972). Clearly, while chance and randomness are intuitive notions, the introduction of an infinite number of events, as reflected in these axioms, will weaken the link with intuition (Feller, 1971; Loeve, 1955) and also lower the commitment of mathematical theories to observable events and real random processes.

It is also unclear to what extent given operational rules suffer from arbitrariness, and Kolmogorov's approach is inadequate or incomplete (Fine, 1973: 59). Various alternative frameworks have been offered: changing the choice of field or assignments of probabilities, weakening tentative connections with relative-frequency notions, recognizing differences in measurability, and questioning characterizations of independence. An obvious first step was to re-express the basic axioms in terms of conditional probabilities. Renyi (1970: 34–5), for example, argues 'the basic notion of probability theory should be the notion of the conditional probability of A under the condition B' for 'every probability is in reality a conditional probability', and the axiomatic scheme ought to reflect this. For most subjectivists too, re-expression with a conditional primitive was inevitable and reduced misunderstanding (de Finetti, 1972; Lindley, 1965: xi). Other mathematicians have opted for expected values as their primitive concept. For example, Whittle (1970) argues that individuals typically understand the nature of an average or mean better than relative frequencies. He also feels that this choice provides a sensible way to link probability with some of its modern applied uses.

Axioms for subjective theories, usually drawing on pairwise comparisons of a type familiar to economists from indifference or utility theory, have generated a great deal of interest – brief lists of major historical references for comparative probability relations are given by Roberts (1979) and Fishburn (1986). The influential treatment by Savage (1954) and other formulations stimulated by potential extensions of the von Neumann–Morgenstern games (as represented in Suppes (1956), for example) require a strong commitment to optimization (the maximization of expected utility) and they raise significant issues in connection with the cardinality of both utility and expectation. It is convenient for us to delay discussion of these issues and some relevant aspects of the corresponding axioms to the next chapter, and to turn instead to a few practical and normative matters which are linked to rational conduct or decisions, coherence and the elucidation of probabilities, including the determination of who (or what) is envisioned to be directly involved with subjective beliefs and other forms of probabilities.

RATIONALITY, COHERENCE AND ELUCIDATION

When probabilities are identified with partial beliefs and individual measures of confidence, simple coordinating principles for the manipulation of probabilities become necessary. These principles have generally been discussed in terms of 'rational' conduct and coherence – fashionable language which promotes the maximization of expected utility and individual consistency with a few axioms for individual preferences and probability measures (often related to economists' treatment of utility and to wider analogies with sensible bets). As a normative guide to rational conduct and the interpretation of evidence, these two elements provide a framework for detecting inconsistencies and recasting numerical values of probabilities. In practice, 'inadequacies' in reasoning (contradictions to the idealized theory) produce some outcomes that may contradict the guiding principles and may yield different assignments of values across individuals with the same evidence.

Any conflict between a given collection of idealized axioms and practical conduct is difficult to resolve, and perhaps need not be resolved. Advice is mixed in this situation. Some observers insist that 'when the axioms appear to clash with real behaviour, one's first reaction should be to attempt to change real behaviour for the better' (Berger, 1986: 351). Others take a less-sanguine attitude to the acceptability of the formal guidelines and look more closely at the acquisition of information, our growing understanding of individual perception and cognition, inattentiveness, complexity, and the nature of the behavioural approximations which are involved in the formal characterization of choice situation – a direct challenge to normative bases.

> ...the perceived world is fantastically different from the 'real' world. The differences involve both omissions and distortions, and arise in both perception and inference.... The decision-maker's model of the world encompasses only a minute fraction of all the relevant characteristics of the real environment, and his inferences extract only a minute fraction of all the information that is present even in his model.
>
> (Simon, 1966: 19)

Various aspects of this conflict are discussed in the next two chapters as we assess the connection of fair bets with rational behaviour, the emergence of game theories, and the distinguishing of substantive rationality from other concepts of rationality.

One issue that should be noted here involves the separation of decision-maker from adviser or observer. Some proponents of the normative framework may understate the significance of this important separation or may treat the decision-makers as less competent to assess their own situation and their options for rational action than the external 'expert' observer. Nor is this lack of hubris limited to subjectivists – advocates of frequency approaches can be equally insensitive to their remoteness from relevant information and to practical obstacles affecting the use of some probabilistic models (including the ability to put off choices, the presence of potential ambiguities, and the desire for enough flexibility to make subsequent changes). Comments on Hamaker (1977) reveal various attitudes to remoteness of statisticians and clients in relation to prior probabilities.

The focus on individual rationality and coherence seems to weaken the application of subjective probabilities to collective economic decisions. Practical guidance to collective rationality is much more complicated. There is a strong temptation to deal only with representative individuals, to avoid major behavioural elements of group dynamics, and to accept 'as if' assumptions of some economic theorists. Savage provides an optimistic illustration:

> It is brought out in economic theory that organizations sometimes behave like individual people, so that a theory originally intended to apply to people may also apply to (or may even apply better to) such units as families, corporations, or nations.
>
> (1972: 8)

There is something disturbing about this popular avoidance of problems with aggregation and consensus formation. This concern widens when computer software is developed to facilitate group decisions, as in expert systems. The 'you' of our normative theory then becomes more complex and the basic intuitive ties to human mental activity are effectively weakened.

> 'You' may be one person, or an android, or a group of people, machines, neural circuits, telepathic fields, spirits, Martians and other beings. One point of the reference to machines is to emphasize that subjective probability need not be associated with metaphysical problems concerning mind.
>
> (Good, 1962: 319)

23

The scope for objectives other than expected-utility maximization and for promotion of some alternative rules to achieve coherence is obviously broadened when the usual focus on individual conduct and judgement is replaced by a (more realistic) collective one. This shift of focus need not weaken support for Bayesian updating of probabilities or for the prior-posterior representations of evidence, but it should attract more attention to the practical aspects of elucidation, dialogue, inductive competence, context and searches for relevant information – even if the normative reference to idealized rationality and coherence is maintained. Some of these aspects are considered in Chapter 6.

Elucidation of probabilities is always a major problem in practical situations. There are always costs involved in finding or estimating probabilities, and there are always imprecision and methodological ambiguities in numerical representations which can point to indeterminacy and potential unreliability. These are awkward features which cannot be simply resolved by invoking familiar statistical criteria such as unbiasedness or the various notions of efficiency. In many respects, the practical need for elucidation in statistical analysis is the counterpart of a need to recognize major transaction costs in several areas of economic activity.

SOME BASIC ISSUES

Economists frequently use terms such as 'information' and 'evidence' without giving sufficient attention to the problem of establishing whether their samples or individual events are 'representative', and so offer an adequate basis for empirical generalization. Frequentists rely on notions of bias, balancing of samples and randomization in the design of experiments and surveys, but selection biases and the distortionary influences of incomplete response patterns create substantial difficulties for the demonstration of representativeness in practice. It is now commonplace to see bold assertions that the poor population is undercounted (overcounted), that economic experience in the Great Depression has been mismeasured, that experimental evidence on impacts of various subsidies are distorted by family instability, that data for aggregate investment fail to deal adequately with technical change (e.g. in connection with development of the computer industry), that durable prices should reflect the pace of depreciation through user-cost adjustments, and that the rapid emergence of service industries has reduced the quality of national

24

accounts. All of these assertions illustrate issues of determining populations, drawing samples, finding appropriate formulae, adjusting for change and other technical matters which affect representativeness and elucidation.

For subjectivists, other problems emerge from the flaws in individual perceptions or judgements of randomness, from overconfidence and from contextual influences (as described by the contributors to Kahneman *et al.* (1982) on the basis of the findings from psychological experimentation and raised in Chapters 4, 5 and 6 below). The pitfalls in intuitive reasoning, limited reliability, common reliance on heuristics and other aspects of numerical calculation encourage us to search for sensible extensions or constructive amendment to probability, such as the advocacy of belief functions by Shafer (1982, 1986) and Dempster (1985), the stress of protocols to guide sequential analysis by Shafer (1985), the use of fuzziness and possibility distributions by Zadeh (1978) and the various proposals for debiasing and improvement of inferential procedures, including recognition of practical non-coherence as discussed by Wilkinson (1977).

Alternatively, we can settle for much less either by using non-unique probabilities and inequalities (e.g. upper and lower bounds) as in the compromise promoted by Good (1962, 1965), or by developing better treatments of ignorance in which probabilistic notions are discarded as unhelpful in particular circumstances (Fine, 1973). These strategies reflect some unfashionable reactions to the overspecification of probabilities, perhaps by moving toward the weakening of arbitrary axioms (to the promotion of qualitative approaches) and to a realistic awareness of the uniqueness and ambiguity of some decisions, or to a flexible recognition that the timing and nature of decisions can often be altered until more evidence is available.

2

EXPECTED UTILITY

The practical elements of gambling and insurance have exerted a strong and persistent influence on the emergence and diffusion of probability models over the last two centuries. For economists, hypothetical gambling and 'games' have also been very influential in connection with the formal treatment of expectations, utility in various forms and optimality or rationality – with 'paradoxes' attracting much attention in the discussion of abstract economic theory and in the resolution of any linkages between theoretical assumptions and the real characteristics of economic choice. In the following sections, probabilistic aspects of this formal treatment are described in five areas of interest, which we identify with the notion of a reasonable or 'fair' bet, the relative measurability or cardinality of individual preferences (as represented by utility), individual aversion or attraction to risk-taking, the fashionable game structure introduced by von Neumann and Morgenstern in the 1940s, and alternative views clarified through paradoxes and other criticisms of some popular ingredients of current economic models.

Contributions to these five areas of interest are often quite remote from any genuinely practical considerations, yet they reveal the inadequate reconciliation of models, their predictions and the reasonableness or realism of their ingredients. Many economists seem uncomfortable when dealing with basic criteria for the choice of theoretical assumptions and abstractions. Their use of expected utility illustrates awkwardness in the integration of probability and preferences in many situations of uncertain choice, the common intrusion of normative elements, a neglect of significant economic considerations (e.g. bankruptcy, credibility, flexibility, learning and novelty), and the willingness of many to postpone an effective clarification of fundamental notions (such as cardinal utility) so that ambiguities persist.

26

Taken in isolation, this general indictment is excessive. In the next three chapters, various attempts to extend and enhance the conventional treatment by economists of choice in uncertain situations are indicated.

FAIR BETS

The best known example of many attempts to clarify the nature of fair bets in uncertain situations is the potential resolution of the St Petersburg Paradox first offered by Bernoulli (1738 [1954]), then endorsed and further developed by Laplace, which focused attention on the relative attractiveness or 'utility' of economic rewards, the interaction of utility with a problem of divergent mean values and the 'paradoxical' conflict between such values and real choices. The history of this paradox (which emerged in the early 1700s) and many of its primary features are described by Jorland (1987), Fellner (1965) and Samuelson (1977). A convenient English translation of Bernoulli's initial paper is often found in books of readings for economists and students of business finance. Daston (1988), Theocharis (1961) and Todhunter (1865) also situate Bernoulli's contribution in historical and philosophical context and reveal its significance within the development of economics, especially mathematical economics.

Although modern economists seldom question the assumption of mathematical expectation (or mean value) as the primary guide to strategic choice among a fixed set of uncertain alternatives, this popular concept of averaging did not emerge until the mid-1650s. As noted by Hacking (1975: 92), the identification of mathematical expectation with fairness or reasonableness was novel then: 'a gambler could notice that one strategy is in Galileo's words "more advantageous" than another but there is a gap between this and the quantitative knowledge of mathematical expectation'. Conjecture remained an art rather than a calculation but, increasingly, prominent authors sought to provide 'better' and more formal rules to guide conduct and choice (Daston, 1988). Discussion of the abstract St Petersburg paradox and similar mathematical puzzles provided a convenient platform for the development, assimilation and promotion of a resilient calculus of probabilities. This new calculus was to apply to basic reckoning and reasonableness, much beyond earlier tentative searches for broad judgemental principles. It widened the previous consideration of 'fair prices' to a new context of rational gambles and inherent uncertainty or riskiness, especially when Bernoulli, Laplace

27

and others stressed the distinction between the mathematical expectation of economic rewards and their 'moral expectation' (by revaluing possible rewards in a gamble according to weights which reflected diminishing marginal utility of income or wealth).

The St Petersburg paradox seems straightforward. Suppose that Peter and Paul are two gamblers who bet on the repeated tosses of a coin. Peter offers to pay Paul an economic reward which depends on when a head first occurs at some toss of the coin, while Paul must pay an initial charge to participate in the gamble. Given a firm schedule of rewards (varying with the timing of a first head) and a known distribution for the probabilities of first occurrence, Bernoulli's puzzle hinges on the definition of a fair price for Paul's participation. Let i (i = 1, 2. . .) indicate particular tosses, a(i) represent the monetary reward to Paul if the first head occurs at the i-th toss, and p(i) be the probability of the first head occurring at that particular toss. Then it is tempting to consider the sum of products a(i)p(i) as a guide in determining the price of Paul's participation. This sum is the mathematical expectation or mean value of the return to Paul before subtraction of his initial payment.

If the gamble continues without constraint until a first head occurs, then some choices of a(i) and p(i) will yield an infinite mean return to Paul. Presumably, this infinite value implies that any unqualified dependence on mathematical expectation provides an inadequate basis for determining the fair (and finite) price for participation. Bernoulli illustrated this mathematical problem of non-convergence by assuming that potential rewards a(i) increase geometrically (doubling at each toss), while their probabilities p(i) decline geometrically (halving at each toss), so their products a(i)p(i) always have a unit value. A paradox emerges here as 'the mathematical theory is apparently directly opposed to the dictates of common sense' (Todhunter, 1865: 220). Bernoulli's solution was to replace a(i) by U[a(i)] so that, with simple constraints on the scaling function U, the revised sum or 'moral expectation' might be finite. When this scaling function is interpreted in terms of cardinal utility, the mathematical expectation of economic returns is thus replaced by their 'expected utility'. Notice that this adjustment requires utility measures to have strong arithmetical properties – compatible with scale multiplication and addition. If such measures are also to yield sensible (finite) values for *all* possible schedules of economic returns a(i), beyond the particular elements of Bernoulli's illustration, then utility itself must be bounded as indicated by Menger in 1934 (Bassett, 1987).

Interpretation of the weights in the scaling function U has typically been associated with variations in the relative value or utility of money to Paul – perhaps unnecessarily so if the a(i) remain small – but clearly required if some potential rewards are sufficiently large to dramatically affect overall wealth. Bernoulli boldly chose a logarithmic scale for the function, klog[a(i)/c] where k and c are arbitrary positive constants (the latter being identified with some 'minimum of existence'). This specification supported a finite moral expectation for his illustration and thus seemed to resolve the paradox for this case. Keynes (1921: 349) suggests that Bernoulli's specification is the first explicit use of the important notion of diminishing marginal utility of money (income or wealth) in the evolution of mathematical economics.

Other realistic adjustments to the structure of the gamble, some not involving an assumption of bounded or diminishing utility, also cause the paradox to disappear. For example, Paul's utility function might be practically constant once the a(i) exceed some threshold. Also, a long series of successive tails may truncate the gamble because the coin or tossing mechanism may no longer be trusted, while an insufficient belief in the solvency of Peter (so that he can make any stipulated payment) can affect the gamble. Further, Paul might not bother to consider very unlikely or very small contributions to mean returns (Fellner, 1965: 107).

Major flaws in generalizations of the St Petersburg puzzle are the common assumption of unlimited capital to participate in the gamble and the absence of any sensible stopping options, which characterize realistic gambling situations. In line with more recent statistical analyses, we have to acknowledge the possibility of gambler's ruin, some potential interest in the size of fluctuations of good fortune, and the normal occurrence of some compulsive or non-result-orientated attitudes to gambling. The recognition of such elements effectively weakens the St Petersburg paradox as a means of clarifying the use of moral expectation or expected utility to guide actual choice in uncertain situations. Taking these elements into account – as, for example, by Feller (1945, 1968) – it becomes clear that any identification of fair price with mean value may be quite unfortunate for the gambler.

Rejecting all random variables with infinite expectations from a fair gamble (as pathological or without practical significance) is also foolish. This was part of a long tradition, or classical theory, in which ignorance and common misunderstandings produced a false sense of importance for infinite values of mathematical formulae, ignoring the existence of finite alternatives for basic measures of location and dispersion, for example, and limiting the character of most acceptable stochastic processes. Similarly, when dealing with individual behaviour or choice, the excessive pre-occupation with expectations and other probability calculations based on long-run convergence ignores the irrelevance and myopia of such ingredients for any unique or casual situations (Hacking, 1975: 94–5) – an illegitimate application even when the inherent probabilities are frequentist. Nevertheless, the reliance on expected utility is quite pervasive in much of current economic theory for abstract choice under uncertainty, while the earlier language of 'fair' bets and 'reasonableness' have gradually been supplanted by a new language connected with 'optimality'. This assimilation occurred without an adequate specification of either computational processes or the costs, origins and operational feasibility of acquiring sufficient information on the facts underlying choice (Simon, 1983: 14).

Before the search for 'reasonable' behaviour is abandoned, we should recognize that the St Petersburg paradox says little about actual choice because its premises are unduly simplified and many relevant features of gambling are ignored. Thus, comparing infinite expected values of economic returns with some vision of commonsense to judge acceptability of theoretical assumptions is distortionary, since the paradox disappears as soon as more realistic adjustments are permitted. The description of an abstract, unrealistic and non-repetitive gamble is hardly a sensible basis for advocating the persistent use of expected utility as either a descriptive or a normative basis for choice. It is simply a convenient means of display, rather like Aesop's fables display important messages to his readers.

A more comprehensive assessment of expected utility is given by Schoemaker (1982), who identifies four different perspectives on the use of this notion – descriptive, predictive, postdictive and prescriptive. He also discusses some evidence and interpretations for validation of economic models which stress the maximization of expected utility. Underlying the alternative views of paradoxical fables as connected with utility, there lies a profound reluctance of economists to address methodological questions in which premises and predictions are (or can be) compared with realistic standards. This troublesome reluctance will re-emerge when we look at the Friedman–Savage treatment of risk aversion and consider the Allais paradox in later sections below and in chapters when substantive or operational significance and rationality are discussed. Basic criteria for the appropriateness of assumptions – including their realistic nature or empirical congruence, apparent fruitfulness and generality for predictions, intuitive attractiveness, technical convenience and self-evident appeal – are generally understated. The feedback from simple paradoxes to any changes in theoretical assumptions remains ambiguous. Economists also seem to have lost sight of the wider conjectural issues and search for principles of reasonable conduct, clearly identified by Daston (1988) and Jorland (1987), out of which the St Petersburg paradox and similar puzzles emerged.

For intellectual history in the context of probability, the long period of interest in the paradox 'led to the substitution of the law of large numbers for the principle of insufficient reason as the foundation of mathematical expectation and it raised the question of the objective or subjective nature of probability depending on whether it applies to single events' (Jorland, 1987: 181). So far, we have not dealt with some obvious issues involved in determining the fundamental character of either cardinal utility or probabilities in forming expected utility. The former is first considered in the next section, while both aspects achieve special importance in the normative or 'neo-Bernoullian' theory of games, major controversies raised by the assault of Allais on the axioms and empirical suitability of this theory, and the challenges from recent psychological research – all of which are briefly discussed below.

We should recognize that economic choices are often not made by individuals, but rather by committees, firms, planning groups, syndicates and households or families. Thus, the interpretation of both utility and (subjective) probabilities is clouded by aggregation, consensus seeking, within-group interactions and dialogue. The meaning and potential measurement of expected utility seem even less transparent when relevant features of the composite economic agents are identified, and they are not enhanced when recast as mythical and single-minded 'representative individuals' so that theoretical models remain tractable but less realistic. Neither Peter nor Paul of the St Petersburg paradox seem representative of composite agents to an extent that the paradox is interesting for a sensible illumination of the principles for composite choice. A contrary view was expressed by Savage, the prominent architect of subjective expected utility (SEU) theory:

It is brought out in economic theory that organizations sometimes behave like individual people, so that a theory originally intended to apply to people may also apply to (or may even apply better to) such units as families, corporations, or nations.

(Savage, 1972: 8)

This perspective seems self-serving and a distortion of perceived benefits from the excessive abstraction in microeconomic theory. We typically ignore the major issues of aggregation not because we should do so, but rather because the difficulties that they entail are so awkward to reduce. In Chapter 6, we consider attempts to deal with aggregation and group decisions.

MORAL EXPECTATION, MEASUREMENT AND CARDINALITY

Utility is associated with desires, wants, satisfaction, worth and preferences. Although not directly measured, most economists routinely use the various concepts of utility and marginal utility – subject to satiation, elasticity, inter-temporal alteration and qualitative constraints such as the *law* or principle of diminishing marginal utility – according to which 'the marginal utility of a thing to anyone diminishes with every increase in the amount of it he already has' (Marshall, 1920: 79). As noted earlier, drawing on intuition and introspection rather than just observation, Bernoulli assumed concavity of numerical utility (as affected by increases in potential income or wealth) in his logarithmic transformation for the St Petersburg paradox.

By the beginning of the present century (through the efforts of Jevons, Menger, Walras and others), a mature theory of utility began to emerge and the issue of its measurability surfaced – preparing the ground for a shift to indifference curves and the ordinal concept of utility (promoted by both mathematicians and economists such as Edgeworth, Pareto, Fisher, Johnson and Slutsky), which became commonplace after being clearly described by Hicks and Allen in 1934. (A standard history of the evolution of utility theories from Smith to about 1915 is provided by Stigler (1950), while the later revisions of cardinal utility and axiomatization based on the von Neumann–Morgenstern framework are clarified by Ellsberg (1954), Samuelson (1950) and Strotz (1953). See also the terse survey of Black (1990).) Despite the shift to indifference curves and the operational advantages of ordinality, the strong discontent with introspection as the primary basis for utility led Fisher (1927) and Frisch (1932) to seek

32

better methods for the direct measurement of marginal utility, but their efforts to find empirical bases for utility were generally unrewarded.

It was left to von Neumann and Morgenstern during the 1940s to produce another concept of cardinal utility, quite different from the traditional one, and thus to challenge the theoretical primacy of the ordinal approaches to demand analysis and to stimulate the experimental validation or measurement of utility (preferences), as characterized by the initial von Neumann–Morgenstern axioms and subsequent reformulations. The newer expected-utility theories are remote from most historical antecedents, despite being labelled as neo-Bernoullian. They produced a very lively debate to clarify how axioms introduced the novel form of cardinality, how personal or subjective probabilities might replace their older frequentist counterparts, how (and how often) the axioms might be violated in practice, whether the new theories are descriptive or normative, what additional benefits could (and would) accrue from the adoption of a game-theoretic structure, whether equilibrium solutions for rational conduct in this formal structure are both determinate and unique (or need the *ad hoc* introduction of additional principles), how solutions might attribute too much information and rationality to economic agents, and whether the structure is consistent with an idealized fully-rational outcome which emerges from some adaptive processes involving learning and bounded rationality.

Such issues produced a radical intellectual climate which is more complex than the tentative comparison of theoretical premises and their potential congruence with vague empirical standards. The menu of economic research was dramatically widened to some areas of mathematics beyond the classical methods of differential calculus – which had provided a standard mathematical apparatus for use in the marginalist revolution (Ingrao and Israel, 1990: 193) – and to other important disciplines, such as psychology, which might offer a better understanding of basic cognitive processes, the stability and firmness of preferences, and the elucidation of probabilities.

RISK AVERSION

The empirical content of expected-utility theory, its validity and implications for economic analyses, and general methodological issues surrounding its potential use were explored by Friedman and Savage

(1948, 1952) in a very supportive way, soon after the second edition of von Neumann and Morgenstern's book appeared. Their explorations reflected the disturbed climate within which economists came to terms with the new concept of cardinality, reformulated and questioned various sets of axioms for expected utility, refined the notions of strategic competition, and explored both parametrization and contingencies in the new theory and earlier alternatives. It soon became clear that there was no solitary set of axioms which uniquely implied maximization of expected utility as the criterion for choice. Rather, several alternatives emerged – including those offered by Marschak (1950), Davidson and Suppes (1956), Herstein and Milnor (1953), Luce and Raiffa (1957), Blackwell and Girshick (1954), and Jensen (1967) – all producing similar representations of consistent preferences with a cardinal utility index. As in the earlier debate over the St Petersburg paradox, it became customary to describe any awkward implications of the axioms as paradoxes.

Friedman and Savage gave their own list of axioms for expected utility and presented a particular function to describe the utility of income. Two properties were especially important:

a) Utility rises with income, i.e. marginal utility of money income *everywhere* positive;

b) It is convex from above below some income, concave between that income and some larger income, and convex for all higher incomes, i.e. diminishing marginal utility of money income for incomes below some income, increasing marginal utility for incomes between that income and some larger income, and diminishing marginal utility of money income for all higher incomes.

(1948: 303)

Later, Savage (1972: 103) noted 'the law of diminishing marginal utility plays no *fundamental* role in the von Neumann–Morgenstern theory of utility, viewed either empirically or normatively ... the possibility is left open that utility as a function of wealth may not be concave, at least in some intervals of wealth'. However, in the influential 1948 account, he and Friedman asserted that 'most consumer units tend to have incomes that place them in the segments of the utility function for which marginal utility of money income diminishes'. This assertion would put most units in the shaded region displayed in Figure 2.1, which reproduces the 'typical shape' of the

34

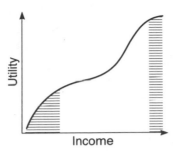

Figure 2.1 Involvement in gambling and insurance

utility curve used by Friedman and Savage (1948) to clarify involvement in gambling and insurance.

Both the 'wiggly utility curve' and the assumed predominance of diminishing marginal utility of income or wealth are commonplace in current microeconomic textbooks, but Friedman and Savage (1948: 297) felt it necessary to justify the descriptive 'realism' of the expected utility hypothesis. Some real problems persist when the hypothesis is applied to gambling and other situations in which the actions of participants are dependent on conditioning events and when the attraction (or entertainment) of participation is not purely result-oriented (Harsanyi, 1978).

By 'realism' or congruence with observable behaviour, Friedman and Savage meant that the assumption of an economic agent acting as if it maximizes expected utility, given a typical utility function, would be compatible with some broad statements (perhaps drawn from casual observations) on the incidence of gambling and insurance in relation to levels of income or wealth. Insurance was viewed as an expression of the choice of certainty in preference to uncertainty, or risk aversion, while willingness to gamble reflected a reversed preference – insurance and gambling offering simple analogies for a host of other and more significant choices (e.g. those involving occupation, line of business activity and investment in securities) in which the potential outcomes can be characterized by income and other contingencies ignored.

The four ingredients of Friedman–Savage (FS) analysis are basic elements (Y, p) representing income (or wealth) of level Y with probability p, the composite bundles combining several of the basic elements $[(Y_1, p_1), (Y_2, p_2), \ldots]$, expected utility of elements $U[(Y, p)]$, and expected utility of bundles $[p_1 U(Y_1) + p_2 U(Y_2) + \ldots]$. Suppose Y_1 represents the initial income of an economic agent, Y_2 is the net income if insurance is purchased, and Y_3 is the net income if a loss occurs which is not covered by a particular insurance payment. Then, if p is the probability of that loss occurring, the economic agent can be viewed as choosing between two alternative situations, namely buying insurance (A) and not buying insurance (B), according to the relative levels of expected utility for the events, EU(A) and EU(B) where EU(B) is $[pU(Y_3) + (1-p)U(Y_1)]$ and EU(A) is $EU(Y_2)$, and where Y_2 lies between Y_1 and Y_3. Clearly, choice based on the EU criterion will be affected by the qualitative features of marginal utility in the relevant range of potential income, with risk aversion and risk attraction being defined in terms of marginal utility.

In Figure 2.2, two alternative parts of the FS utility curve are shown. These parts (i) and (ii) correspond to diminishing and non-diminishing marginal utility. Other lines (iii) and (iv), linear segments, are generated for EU(B) as p varies between 0 and 1 in the two

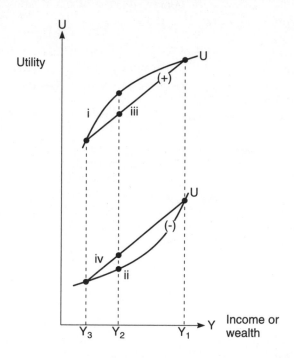

Figure 2.2 Two alternative parts of the Friedman–Savage utility curve.

Note: Lines (iii) and (iv) represent EU associated with different probabilities of uninsured loss.

cases. The vertical line at income Y_2 has intersections with these lines to generate two 'gaps' for [EU(A) - EU(B)], one positive and one negative, determined by the patterns for marginal utility. The attractiveness of insurance depends on the sign of the gaps. For an analysis of gambling, primarily a simple lottery, Y_2 can be considered income prior to purchase of a lottery ticket, while Y_3 and Y_1 represent the potential income associated with not winning or winning the lottery, respectively. The situations A and B can then be identified with abstention (keep Y_2, certainty) and participation in gambling, but the relative size of the intervals on the horizon axis must be changed to reflect various conditions for gambles. Again the FS analysis is reasonably straightforward if the marginal utility of an individual agent can be placed within the pattern of their wobbly curve. More general propositions on the incidence of insurance and risk-taking emerge as the population of economic agents is assigned to the three broad concave or convex portions of their utility function. Challenges to FS propositions and the recognition of some additional ones can also be expressed by advocating the use of a different shape for marginal utility, as illustrated by Markowitz (1952a).

GAMES, UTILITY AND RATIONALITY

In our brief account of the Friedman–Savage papers, we failed to indicate how probabilities were discovered and gave inadequate attention to the meaning and measurability of their basic concepts of consistent preferences and cardinal utility. Calls for better terminology to deal with the various meanings of utility go back at least to Fisher (1918) and they became more pronounced as the index of von Neumann and Morgenstern was assimilated by economists, and as Savage and other statisticians gave more attention to the subjective source of probabilities in expected-utility theories. The Neumann–Morgenstern (NM) index of cardinal utility and its properties emerged from the acceptance of various axioms, which persisted so long as they were perceived to be plausible or compatible with broad predictions of the type noted by Friedman and Savage. Since the NM index was different from earlier notions of cardinal utility, another term should have been introduced (Ellsberg, 1954) – not least because the new index was recognized as arbitrary within a class of linear representations, while choice in the absence of uncertainty was (and still is) being treated within an ordinal framework. Ozga indicates the general background and status of the NM index:

In Bernoulli's theory the subjective pay-offs are in terms of the 'intrinsic value' of what the individual may win. Marshall's utility is the degree of satisfaction ... from the objective pay-offs. In either case the relevant term describes what the individual would experience or what would be his state of mind if the respective outcome came off . . .. But the Von Neumann and Morgenstern's utility is not a cardinal measure of value or of satisfaction. The fact that the utility of a ... pay-off is twice as large as that of some other does not mean that the satisfaction from getting the former is twice as large . . .. Utility in this case is only a theoretical term which has no other significance than that of the result of a particular operation.

(Ozga, 1965: 188)

The arbitrariness of the new index is simple to demonstrate. Suppose we consider choice between the certainty of $600, represented by A:(600, 1), and a bundle of $500 or $1000 with probabilities c and $(1-c)$, B:[(500, c), (1000, $1-c$)]. There will exist some value of c, say c^*, such that the individual agent is indifferent between the two alternatives, say $U(A/c = c^*)$ is equal to $U(B/c = c^*)$. From expected-utility theory, we would expect any acceptable utility index UI to satisfy the equation:

$$UI(600) = c^*UI(500) + (1-c^*)UI(1000).$$

If UI(500) and UI(1000) are given arbitrary values (e.g. 0 and 1), then a value can be attached to UI(600), namely $(1-c^*)$. More generally, if N is any value between 500 dollars and 1000 dollars, there will exist some value c such that the agent is indifferent between (N, 1) and [(500, c), (1000, $1-c$)] and thus we require:

$$UI(N) = cUI(500) + (1-c)UI(1000),$$

so UI(N) is $(1-c)$, given the arbitrary specifications for the two reference points. On the other hand, if M is any value higher than 1000 dollars, there exists some value m such that we would require:

$$mUI(500) + (1-m)UI(M) = UI(1000),$$

and so UI(M) is $1/(1-m)$. Changing the reference points for the NM index will change the values of UI(N) and UI(M) but will not affect the application of the index in expected-utility theory.

For any income Y, there are a multitude of alternative values, UI(Y), made unique by specifying the arbitrary values of any two particular

38

points in the scale – i.e. alternative choices for UI(Y) are linearly connected so that specifying the two arbitrary reference points is equivalent to choosing among alternative linear transformations. Specifically, if UI(Y) and UI(Y) are alternative indices, then UI(Y) is equal to $[e + f$UI(Y)$]$ for some values of e and f. With any income levels N and M, the numerical 'values' of the ratio and difference of the corresponding utility indexes, UI(N)/UI(M) and [UI(N) - UI(M)], differ according to the anchoring specification – rather like familiar measurements of temperature. Any UI(Y) is not to be confused with the 'utility' of UI(Y). Note also that there will exist (within EU theory) an equivalent certain income, D dollars say, for which the individual is indifferent to D:[$(Y_1, c), (Y_2, 1-c)$], a composite bundle with given incomes Y_1 and Y_2, and probability c. The extension of EU theory to larger bundles with three or more components is obvious.

This account is deceptively simple as various assumptions have not been made explicit. Usually, the specification of a few well-chosen axioms will remove any imprecision, enabling the 'realism' of the theory to be compared with the plausibility of these axioms and also encouraging researchers to seek alternative, perhaps less onerous, specifications. Over time, the standard of plausibility has been partially overtaken by a promotion of normative views – with EU axioms representing what individual agents should do rather than providing an empirical framework to describe how they appear to act. The term 'rationality' has thus come to represent both the hypothetical consistency of choice with a set of arbitrary axioms or principles and the normative appraisal of actual choice.

For notational convenience, let U(A) and U(B) represent some given preferences in the sense that U(A) is set equal to U(B) when the individual is indifferent between prospects A and B, and we use the symbols $>$ and $\geq$ to indicate that the prospect A is preferred to B, [U(A) $>$ U(B)], or that A is preferred at least as much as B, [U(A) $\geq$ U(B)]. This notation can be used to describe some of the axioms which have been proposed as the bases for expected-utility theories, characterizing patterns of consistent preferences. It is assumed that any rational choice is defined over a given collection of alternative choices or prospects, which can involve comparable quantities of goods, lotteries and other mixed bundles. This collection of alternative prospects and their ordering according to preference or indifference are assumed to be closed and completely ordered:

1 for any prospects, e.g. A and B, either $U(A) \geq U(B)$ or $U(B) \geq U(A)$, or both (when there is indifference between A and B);

2 if all preferences are transitive – 'a plausible and generally accepted property' of preferences according to von Neumann and Morgenstern (1944), but later questioned by some economists and psychologists from experimental evidence – in the sense that if $U(A) \geq U(B)$ and $U(B) \geq U(C)$ for any prospect C, then $U(A) \geq U(C)$; and

3 if A and B are prospects, then so too is $C:[(A, p), (B, 1-p)]$ for all scalars p between 0 and 1.

We might also require that:

4 $U[(A, p), (B, 1-p)]$ should lie between $U(A)$ and $U(B)$ if and only if $U(A) \geq U(B)$, for all such ordered A and B.

Other potential axioms include that of 'independence', whereby:

5 $U(A) \geq U(B)$ if and only if $U[(A, p), (C, 1-p)] \geq U[(B, p), (C, 1-p)]$ for all scalars p between 0 and 1, and for any A, B and C; and

6 for all A, B and C such that $U(A) \geq U(B) \geq U(C)$, there exists some p so $U(B)$ equals $U[(A, p), (C, 1-p)]$, the 'Archimedean axiom'.

Other common requirements deal with the distribution of probabilities among prospects and some technical matters (e.g. the existence of enough non-indifferent prospects). Finally, the fulfilment of a sufficient batch of axioms leads to the von Neumann–Morgenstern result that it is appropriate to replace our trivial notation U(.) by a scale (monotonically increasing) function UI(.) with definite mathematical properties, which can be used in the simple manner of Friedman and Savage illustrated above – where maximization of a corresponding expected utility E[UI(.)] yields rational choice and a sensible characterization of preferences. From the EU theory, when A, B and C are cash rewards, the independence and Archimedean axioms become requirements on the UI function that:

pUI(A) + $(1-p)$UI(C) > pUI(B) + $(1-p)$UI(C) if and only if A > B, for all scalar p where $0 < p < 1$;

there exists some p such that if UI(A) > UI(B) > UI(C), then UI(B) = pUI(A) + $(1-p)$UI(C).

Clearly, if conditioning events and non-result-orientated influences do not substantially affect choice, then a tentative validation of these axioms as applicable to actual choices can be sought with the design of experiments and questionnaires which present appropriate mixed choices. Unfortunately for the EU theory, as reported in a later chapter, the results of such approaches offer a wider vision of decision frameworks and of influences on choice, often remote from the pristine simplicity of the conventional theory but more 'realistic' in recognizing various cognitive and framing 'biases' that can occur and the significance of major contextual factors. (See also the pessimistic judgements expressed by Fisher (1989) and Sutton (1990) on the limited explanatory power and narrow domain of applications for game theory, even after its four active decades of development.)

Empirical validation is not a simple matter as any judgement of particular axioms can depend on many logical, methodological and practical considerations – some of which were identified by von Neumann and Morgenstern (1944):

> A choice of axioms is not a purely objective task. It is usually expected to achieve some definite aim – some theorem or theorems are to be derivable . . . – and to this extent the problem is exact and objective. But . . . there are always other important desiderata of a less exact nature: the axioms should not be too numerous, their system is to be as simple and transparent as possible, and each axiom should have an immediate intuitive meaning by which its appropriateness may be judged directly.

The importance of desiderata remains an important area of debate. Howard (1992), a prominent decision analyst and supporter of the 'old time religion', provides a convenient list of them (summarized by Eppel et al.), while noting the inevitable myopia of EU theory and its rivals:

> The decision problem . . . should be guided by the objectives to be achieved by the analysis and by the selection of axioms for that analysis. Indeed, the axioms of expected utility, or a competing set of axioms, provide only part of the . . . assumptions necessary for prescriptive decision analysis.
>
> (Howard, 1992: 274)

Experimental results and responses to questionnaires may not adequately reflect the particular circumstances of any choices or decisions. They can, however, suggest new lines of development for generalization of EU theory and indicate some constraints on the situations to which this theory and its rivals might be applied (e.g. excluding large payoffs for lotteries and also prospects with low probabilities). We look at the generalization of EU theory and some experimental violations of the standard axioms (independence and transitivity) in Chapter 6. How modified theories are to be validated remains unclear, but economists will probably continue to muddle along with a mixture of empirical and normative arguments. Already, reflecting the fashionable use of mathematics as clarified by Debreu (1986), there are suggestions that 'any convincing changes will have to be made at the level of the axioms' (MacCrimmon and Larsson, 1979). Both von Neumann and Morgenstern believed that the axiomatic method could (and should) be applied to all sciences, including economics, provided they are sufficiently developed – the method being indispensable for the clear construction of scientific theory (Ingrao and Israel, 1990: 192). Obviously, their treatment of games and expected utility provided a strong stimulus to the fostering of axiomatic models in economics.

Almost certainly, the EU theory will be retained (somewhat tarnished) despite any empirical failures – 'as we have learned throughout the 1980s, most results obtained using game theory are delicately dependent on the details of the model (extensive form) employed, and many of these details are impossible to identify or proxy empirically' (Sutton, 1993) – through the mental jump to normative standards for the resolution of 'irrational' conduct and through the optimistic promotion of benefits which might accrue from the sensible instruction of participants in choice or decision making. This development is clearly illustrated by Morgenstern, who supports the EU theory and NM axioms as normative in a sense that:

> the theory is 'absolutely convincing' which implies that men will act accordingly. If they deviate from the theory, an explanation ... will cause them to readjust their behaviour ... Naturally, it is assumed that the individuals are accessible intellectually .... In that sense there is a limitation since there are certainly persons for whom this is impossible. Whether they then should be called 'irrational' is a matter of taste.

> (1979: 180)

Other rationalizations add further elements. Harsanyi, for example, introduces a naive notion of responsible conduct:

> it is natural to expect that, in making important policy decisions, responsible decision makers will take a result-oriented attitude toward risk taking. This is probably a reason-

ably realistic descriptive prediction; and it is certainly an obvious normative rationality constraint as well as a moral requirement: responsible business executives using their shareholders' money, and responsible political leaders acting on behalf of their constituents, are expected to do their utmost to achieve the best possible results .... Even clearer is the obligation of taking a purely result-oriented attitude in making important moral decisions.

(Harsanyi, 1978: 224)

However, the intrusion of ethics, morality and responsible conduct of businessmen and politicians in the promotion of normative EU (or SEU) theory seems to add complications rather than offering any better justification for its retention – not least because it seems so remote from the motivations of executives and politicians with whom we are familiar. Harsanyi's unqualified general claim that SEU axioms are 'absolutely inescapable criteria of rationality for policy decisions' is clearly just outrageous nonsense.

Another, more recent, appraisal of the merits and limitations of the EU theory by Dreze (1974) focuses on some perceptions of its relevance, usefulness and generality, and on its integration with other theories. This appraisal also allows a strong normative appeal and the inertial influence of our immanent standards to overcome other scruples, before concluding that expected utility provides a good starting point for some (but not all) economic analyses:

Specific applications may call for further specifications or for specific extensions; ... one should remain conscious of the doubtful descriptive realism .... But that being understood, it seems fruitful to investigate economic models relying, explicitly or implicitly, upon the axioms or conclusions of the theory.

Against this and similar views should be set the aggressive stance of Allais who insists that, for forty years, 'the supporters of the neo-Bernoulli formulation have exerted a dogmatic and intolerant, powerful and tyrannical domination' over discussion in this context (1990: 8). A valuable and much wider checklist of potential 'fatal flaws' for all rational choice models has been provided by Bunge who argues that they are to be found wanting 'for being trapped within the individualistic and utilitarian dogmas, for involving fuzzy basic notions and untestable assumptions, for idealizing the free market, and for failing to match reality'. His list includes the following re-phrased items:

43

(a) they rarely define the subjective utility functions so they are often pseudomathematical despite their symbolic apparatus;

(b) they involve the concept of subjective rationality which serves to save the rationality postulate from refutation;

(c) they are static, assuming the fixity of options and constancy of preferences;

(d) although supposedly individualistic, they smuggle in unanalysed holistic notions such as the situation, market, and institutional framework;

(e) they overlook the power of groups and systems which distort preferences and restrict the freedom of choice;

(f) they ignore planning;

(g) they only cover one type of social interaction, namely exchange;

(h) they omit sources of human action other than self-interested calculation;

(i) they do not solve the problem of micro–macro relations;

(j) they have almost no predictive power because of their excessive generality; and

(k) being unrealistic, at times surrealistic, they are not suitable policy tools.

(Bunge, 1991: 83)

Both Allais and Bunge also dislike the common integration of subjective or personalistic probability in expected-utility models, which followed the development and wide assimilation of additional axioms for consistency of their probabilistic ingredients. Here, normative rationality also came to be attached to the probability axioms, as illustrated by Savage:

the role of the mathematical theory of probability is to enable the person ... to detect inconsistencies in his own real or envisaged behaviour. It is ... understood that, having detected an inconsistency, he will remove it.

(Savage, 1972: 57)

Although partially anticipated by Ramsey (1926) and de Finetti

(1937), subjective expected utility (SEU) theory was launched by Savage (1954, 1972), who enlarged the NM framework to include personal probabilities rather than frequentist ones. Variants of his treatment and related matters are described by Luce and Raiffa (1957), Fellner (1965), Fishburn (1970), Lindley (1971), Roberts (1979) and many others, while Shafer (1988) offers an alternative 'constructive' perspective to repudiate 'the claim that subjective expected utility provides a uniquely normative way of constructing decisions' and to question isolation of normative interpretations for basic postulates from the sensible requirement that they have some empirical content. Analogous to the NM approach to utility, the axioms of Savage are aimed at the simultaneous derivation of basic probability measures to characterize incomplete information or beliefs.

Expected-utility theory is not simply part of game theory, but it was first created as a pillar for game theory (Luce and Raiffa, 1957: 12) and thus it clearly reflects a dramatic shift in economic metaphor away from the 'neoclassical world composed of myriad individual Robinson Crusoes existing in isolation and facing fixed parameters', and towards a neo-institutional framework with fewer participants and strategic choices (Schotter, 1992). Von Neumann and Morgenstern (1953: 28) acknowledged the subordinate role for expected utility – conceding they 'practically defined numerical utility as being that for which the calculus of mathematical expectations is legitimate', initially just to produce a simple theorem on two-person, zero-sum games. With the emergence of game theory, the parametrization of economic choice changed radically, interdependence became recognized to a limited extent, and rational solutions were deduced for a host of special circumstances, though not without some resistance, e.g. Kaysen (1946–1947).

The shift in metaphor attracted more attention to the presence of 'significant others' in bargaining situations and stimulated a fuller (but still incomplete) classification of relative ignorance, promoting:

- adoption of various procedures or rational decision strategies to deal with potential responses of adversaries in the context of uncertainty;
- tabulation of alternative payoffs;
- characterization of options by extensive and normal forms (Raiffa, 1968); and
- search for useful theorems identifying equilibrium states, their existence and uniqueness.

(See, for example, the comprehensive accounts of Borch (1968), Harsanyi (1977), Ordeshook (1986) or Thomas (1984).) However, game theory as a whole does not seem to have extended many economists' appreciation of the wide range of probabilistic notions, unfortunately being mired in the familiar but excessive constraints which have been used to develop both the conventional expected-utility theory and any current extensions of it. Subjective or Bayesian probabilities took on a life of their own, much beyond game theory but normally connected to expected utility and hypothetical lotteries or fair bets (Good, 1990), and have acquired a considerable following among economists and statisticians.

PARADOXES

In large part, the rejection by Allais of the neo-Bernoullian or expected-utility theory took the same form as debate surrounding the St Petersburg paradox — where the implications of an abstract model are held to be inconsistent with observed behaviour in some instances, thus the general applicability of one or more of its postulates becomes questionable. Using notation introduced in the last section, consider the preferences among four alternative situations in the EU context,

A:(100, 1),   B:[(500, 0.1), (100, 0.89), (0, 0.01)],
C:[(100, 0.11), (0, 0.89)], D:[(500, 0.1), (0, 0.9)];

then

$E[UI(A)] = UI(100)$, $E[UI(B)] = 0.1UI(500) + 0.89UI(100)$,
$E[UI(C)] = 0.11UI(100)$, $E[UI(D)] = 0.1UI(500)$.

Subtraction reveals that the difference $E[UI(A)] - E[UI(B)]$ is the same as $E[UI(C)] - E[UI(D)]$, so expected-utility theory implies that if A is preferred to B, then C must be preferred to D, and vice versa. Evidence suggesting the coexistence of both A being preferred to B and C being preferred to D produces a 'paradox' in the traditional sense, within which a combination of the NM index with the maximization of expected utility clearly yields an inadequate representation of real preferences — or else, in the normative interpretation, indicates that preferences should be modified.

For paradoxical situations of this type, Savage illustrates the latter:

a person who has tentatively accepted a normative theory must conscientiously study situations in which the theory seems to lead him astray; he must decide for each by reflection — deduction will typically be of little relevance — whether to retain his initial impression of the situation or to accept the implications of the theory for it.

(Savage, 1972: 102)

Consider another situation in which a fair die is tossed and two alternative schedules for winnings (A and B) are offered to a player, displayed in Table 2.1:

Table 2.1 Schedules for potential winnings (in dollars)

Die:	1	2	3	4	5	6
A	6	7	8	9	10	5
B	5	6	7	8	9	10

Source: Howard (1992: 28)

The two schedules indicate that choice between them involves the same expected utility, but the player might notice that a higher value occurs five times out of six if A is chosen and may be tempted to prefer this option. Now suppose the potential winnings are enhanced so they are expressed in thousands of dollars rather than in smaller units. The difference between winnings if a die shows a six becomes $5000, so the player may now opt for B on the grounds that he would greatly regret choosing A if this particular face of the die appeared. Clearly, in both situations, the EU theory is violated by the potential rejection of indifference, while the realistic explanations of his choices by the player do not seem particularly foolish. Should we acknowledge the EU theory's limitations or assume that the player needs to be offered advice so as to conform to the theory? This question gets to the heart of the normative issue for EU theory, where desiderata become especially important (Howard, 1992).

Allais presented his forceful criticism of the NM approach at a conference on mathematical economics and risk in 1952, but an English translation of this criticism was not generally available until twenty-five years later — when issued with additional experimental results and the presentation of contemporary views in Allais and Hagen (1979). In his latest commentary, Allais insists that neither the St Petersburg nor the Allais paradox are really 'paradoxes' (anomalies, errors, signs of incompetence and inexperience), since the axioms for EU theory are often violated and thus are not evidently correct.

Both correspond to basic psychological realities: the non-identity of monetary and psychological values and the importance of the distribution of cardinal utility about its average value.

The Allais paradox does not reduce to a mere counter-example of purely anecdotal value based on errors of judgement
.... It is fundamentally an illustration of the need to take account not only of the mathematical expectation of cardinal utility, but also of its distribution as a whole about its average, basic elements characterizing the psychology of risk.

(Allais, 1990: 8)

Objecting to the independence axiom, he recognized a 'certainty effect' which led to the suggestion that individual attitudes towards risk are not simply exhibited by the curvature of utility functions, but rather they change for a given individual between different patterns of risk:

I viewed the principle of independence as incompatible with the preference for security in the neighbourhood of certainty shown by every subject and which is reflected by the elimination of all strategies implying a non-negligible probability of ruin, and by a preference for security in the neighbourhood of certainty when dealing with sums that are large in relation to the subject's capital.

(ibid.)

Unfortunately, for some observers, the impact of this interesting effect and other complications is to disturb the convenience of our familiar axioms, without providing anything but *ad hoc* replacements which lack generality. However, this complaint simply underlines the need, already acknowledged by Morgenstern (1979: 177), that all theories be limited to their proper domains – a requirement which allows ambiguous information to provide another basis for leaving the EU framework in some circumstances.

Suppose we consider an individual who has the chance to bet on the colour of a ball, red or black, taken at random from one of two urns. Let [R, I] and [B, I] represent bets on red or black for any ball taken from the first urn, in which the colour ratio is unknown to the potential gambler. Similarly, let [R, II] and [B, II] indicate bets on red or black for a ball taken from the second urn, in which odds of the colours are even. If the gambler is successful, he wins $10, otherwise he wins nothing. Then, we can use the simple notation used above to write U[R, I] and its obvious counterparts as ingredients in illustrating the preference for one bet rather than other bets. Ellsberg (1961) noted that the occurrence of common indifference so U[R, I] equals U[B, I] and U[R, II] equals U[B, II]. However, he also found that both

$$U[R, II] > U[R, I] \text{ and } U[B, II] > U[B, I],$$

providing a violation of familiar EU axioms – that is, the trivial notation U(.) cannot be replaced by any UI(.) index. This result may be interpreted – e.g. by Einhorn and Hogarth (1985), Sherman (1974), Bernasconi and Loomes (1992), and Ellsberg (1961) – as reflecting a distinct 'aversion to ambiguity', although other factors such as 'scepticism' (Kadane, 1992) might also be active when the intrinsic beliefs or probabilities are not sufficiently clear to offer any firm support to numerical requirements.

More generally, Ellsberg points to a group of situations in which (due to the ambiguities, imprecision, and unreliability of information), the relative lack of firmness in a decision framework may be recognized and any estimates of subjective probabilities are seen as inadequate for a fruitful implementation of EU theory. In practice, he notes that decision makers act in deliberate conflict with the axioms of subjective expected utility 'without apology, because it seems to them the sensible way to behave' and asks if they are clearly mistaken. After three decades of the presentation of his paradox, we should acknowledge the commonsense in Ellsberg's two modest suggestions that we avoid EU axioms in the 'certain, specifiable circumstances where they do not seem acceptable' and that we should, in these circumstances, give more attention to alternative decision rules and non-probabilistic descriptions of uncertainty.

Where does this leave us in regard to expected utility? A definite response to this question has to be delayed until other visions of rationality are identified and assessed, some of the principal extensions to (and adjustments of) EU theories have been sketched, and the scope, validity and implications of experimental evidence have been noted. Certainly, prominent opponents such as Allais have ample opportunities to offer constructive criticism – see the contributions to Stigum and Wenstop (1983) and Hagen and Wenstop (1984) – and an active search for alternative perspectives on rational choice is under way, without diminishing the apparent attractiveness of EU theory to many economists.

3

RATIONALITY AND CONDITIONAL EXPECTATIONS

Economists have used the terms 'rational' and 'rationality' for so many different concepts that it is quite tempting to attach subscripts to separate them! Besides the familiar rational choice, for example, we readily find rational degrees of belief (Keynes), rational conduct (Knight), rational processes (Hayek), and rational expectations (Muth, Lucas); even rational termination (Brams) and rational distributed lags (Jorgenson). Similarly, as conveniently summarized by March (1978), there are bounded rationality (Simon, Lindblom, Radner), contextual rationality (March), game rationality (Brams, Harsanyi and Selton), process rationality (Edelman, Cohen and March), adaptive or experiential rationality (Day and Groves), selected rationality (Nelson and Winter), posterior rationality (Hirschman), and quasi-rationality – enough concepts to make one dizzy.

For our purposes, it seems appropriate to concentrate on a shorter list, within which we recognize 'rationality' applied to the internal consistency of axiomatic models and to the pursuit of self-interest through optimization in the conventional mathematical treatments of rational choice; and to the wider behavioural (or procedural) approach of Simon and others, which is associated more closely with cognitive psychology, pattern recognition, rational adaptation and computational limitations. Expectations, especially conditional expectations, of stochastic variables are common for all of these forms of rationality – in basic axioms determining any standard and generalized expected-utility theories, in the goals for normative mathematical models, and in some approximations (involving certainty equivalents) which illustrate heuristics and clarify the realistic limits for individual optimization in a wider microeconomic context.

To explore the troublesome aspects of reconciling probability with rationality in economics, the following sections take up some issues connected to: (certainty) equivalents; portfolio models which force uncertainty into a common two-parameter characterization or mean-variance framework; the extraordinary commitment to rational expectations with ergodic and anti-historical attributes (and with a disturbingly naive treatment of information and aggregation); theoretical attempts to represent the notion of increasing risk in terms of the relative curvature of utility functions; and radical perspectives which draw on direct observation and assume that any psychological laws (e.g. as affecting both preferences and subjective probabilities) are 'limited in scope and generality, and will be mainly qualitative' and their 'invariants are and will be of the kinds that are appropriate to adaptive systems' (Simon, 1990: 2). Covering such diverse areas, our terse discussion is inevitably incomplete (perhaps, partisan and unfair), but it simply offers an appreciation that there are important issues which need to be addressed in this context.

As in earlier chapters, sufficient references to other primary sources are given to encourage a quick access to more detailed and comprehensive treatments. We offer a sampling and indicate the wide scope of major concerns (always with uncertainty, risk and probabilistic elements in mind), but there are important omissions at which we can only hint, most notably in three important areas of economic applications: general equilibrium models with stochastic ingredients and learning, experimental validation of rationality, and the various aspects of game theory which are connected with some newer approaches to equilibrium solutions. Associated with the many forms of rationality, there correspond distinctive non-fulfilment or irrationalities, which differ in relative severity and are affected to various degrees by the mixture of realistic and normative standards noted in the last chapter. Further, there will always be hazards of using and interpreting the technical language of rationality as if it were free from the influence of everyday usage, such as consistent or well-conceived reasoning. It may not be too easy to dismiss something as technically irrational, for example, when it appears otherwise to the lay audience grounded in a different meaning of irrationality.

Internal consistency seems quite straightforward, merely a mathematical condition to be verified and an obvious requirement for any theory. Such consistency does not normally require the acceptance of any ingredients of a particular model, either in the sense of a useful correspondence of its symbols and operations with real magnitudes and processes or in the sense of being worthwhile. However, if internal consistency is defined to include optimality conditions (such as the maximization of expected utility or some other expression of myopic self-interest) and promotes normative implications, then some awkwardness occurs in separating rational from irrational – the combination of consistent axioms and particular goals is adversely affected by common weaknesses in characterizing both mental activities and the economic context.

There is also a problem in which 'backward spillovers' link normative aspirations to (internally consistent) model ingredients. For example, calls for substantive rationality may be weakened in the absence of consistent axioms which fulfil additional (perhaps implicitly desirable) features such as those ensuring the existence, uniqueness, and stability of equilibrium solutions. Consistency (and its probabilistic counterpart, coherence) is then a cloak which obscures the true nature of model construction and potential correspondence with the economic context. Ambiguities and problems may appear to be resolved by a mis-specification of rationality (Sen, 1986), and by the token identification of rationality with unjustified assumptions.

We have allowed reasoning and practical correspondence to intrude in the discussion of rationality as internal consistency – odd behaviour if such consistency is not substantially affected by any 'external' motives – and have acknowledged the intermingling of internal consistency and self-interest in many optimization models. Substantive rationality, where goals are achieved within the limits set by fixed conditions and constraints, is not markedly affected by the intermingling of dual requirements for internal consistency and goal-seeking behaviour or choice. However, the value attached to substantive rationality in normative models may be effectively reduced by any major failures to demonstrate clear correspondence with the economic context and by an inadequate treatment of human reasoning. (What can one make, for example, of the fashionable rational-expectations models which are internally consistent, but permit market-level 'subjective' probabilities and their objective counterparts to be constantly equated, require all variables to have stable moving-average representations, perhaps involve some mindless 'representative' individual, and assume that relevant information is either costless or not wasted?)

Clearly, while internal inconsistency is undesirable, any axiomatic adjustments used to produce consistency and coherence must also reflect both the purposes of modelling and the contextual scope for normative implications. This view of rationality seems to be the most important lesson drawn from the recent research on 'paradoxes', generalizations of expected-utility theories, and the perceived 'violations' of conventional EU axioms. There is a need for more explicit justification in terms of the domain and rules of application, which is often ill-expressed by requiring assumptions to be more realistic. As indicated by Good:

> The abstract theory can be regarded as a machine, a black box that manipulates symbols in accordance with the axioms. There should be precise rules of application that give meaning to the undefined words and symbols. These ... are rules for feeding judgments into the machine and for interpreting the output from the machine as new ... 'discernments'. .... The adequacy of the abstract theory cannot be judged until the rules of application have been formulated.
>
> (1957: 21)

Thus, any rationality in the sense of internal consistency alone is insufficient to warrant much attention, as it trivializes the need for basic axioms to be judged by wider standards and inhibits the interpretation of any model's output.

Turning to internal consistency in situations of uncertainty, there will seldom be a uniquely correct way to express degrees of belief or personal confidence. The subjective probabilities (as used by personalists) are required to be appropriately coherent; i.e. the confidence of an individual is distributed in a way which strictly conforms with the standard axioms of probability, and rationality is directly associated with the presence of such coherence. Since Ramsey (1926), and especially since Savage (1954), it has become customary to think of coherence in terms of avoiding a 'Dutch book' – a situation in which the gambler must always lose because of an inappropriate assignment of subjective probabilities or beliefs. Rationality is expressed here in terms of rational or coherent *beliefs*, and there are preoccupations with normative rules for actions, including avoidance of behavioural inconsistencies, and measurement. However, neither Ramsey nor Savage gave adequate attention to how such beliefs enter the mind, how their potential imprecision may inhibit pair-wise comparisons underlying rational choice, or how they are modified – these are matters that came to be addressed by the proponents of procedural rationality, but some Bayesian statisticians also hint at frequent excesses in relation to the presumption of sufficient precision.

Although subjective theories seem to assign firm numbers to beliefs, this spurious precision may 'be taken with a large pinch of salt' (Smith, 1984: 248) in practice, since the legitimacy of prior and posterior probabilities can always be challenged, giving access to the plurality of any coherent responses to information or data. Good describes this deceptive precision as part of a practical compromise:

> the . . . Bayesian can assume that the probabilities are precise and satisfy the usual numerical axioms, but he holds in mind that the set of all probabilities is not at all unique. He makes judgments of . . . inequalities and infers new probability inequalities with the help of the mathematical theory. He also explicitly recognizes that there is in each application an optimal amount of self-interrogation that is worthwhile.
>
> (1965: 10)

On the other hand, this practical qualification of results-based subjective probabilities may not be widespread. Drawing on everyday language for rationality, Fine points out that Bayesian statistics

> can provide a false sense of security. It is neither rational nor wise to force what few crumbs of information we may possess about a parameter into a misleadingly detailed form of a distribution. It is a seeming defect of most . . . theories of subjective probability that they do not distinguish between those prior distributions arrived at after much experience and those based on only the vaguest forms of uninformed introspection.
>
> (Fine, 1973: 231)

Further, taking into account other contextual omissions, he insists that 'while subjective probability claims to be a part of a normative theory of decision-making, and not a descriptive theory of psychological processes, it can deviate greatly from a descriptive theory only at the peril of being inapplicable' (ibid.: 234).

Three other dynamic aspects of rational decision making have often been neglected by economic theorists (and they are ignored below). First, we should recognize that any current anticipations, beliefs and probabilities may not persist as conditions change and as uncertainties are further resolved – so we might choose to add sufficient flexibility to our future plans, permitting unexpected favourable options to be utilized or any unfavourable ones to be avoided. That is, a sensible current choice of action may require the inclusion of contingencies, hedging, and discretionary 'slack' to deal effectively with the future acquisition of information (perhaps unsought) and changing circumstances. Surprisingly, Hart (1942, 1947) alerted economists to the common need for flexible planning more than fifty years ago. Generally, however, his perceptive comments received insufficient attention until a gradual emergence of heuristic-based approaches to behaviour and learning in the 1970s, although similar notions are now a standard part of some descriptive theories for large firms or conglomerates and are compatible with adaptive microeconomic models.

Part of the neglect of flexibility stems from the additional mathematical complexity that it entails. Hart, for example, noted the use of functionals for subjective uncertainty, which describe the probabilities (or 'likelihoods' of probabilities) as first explored by Tintner (1941). A more modest adjustment to theoretical models involves giving attention to the higher moments of a single estimated distribution of future returns (beyond expected means), partially illustrated by variances or standard errors in portfolio models, while much more radical approaches are found in the drastic simplification of potential choices and partial abandonment of some probabilities in the Carter–Egerton theory for business decisions (Carter, 1972: 40), or in the complete abandonment of numerically precise probability judgements.

A second area of benign neglect involves the timing of choice or action, both of which can be deferred until more information is available and corresponding risks have diminished. Further delay permits an expansionary search for (and inspection of) additional options, lengthier deliberation, access to other external sources of information, the reduction of uncertainty to acceptable levels, some resolution of ambiguities, the withdrawal or weakening of any opposition in group decisions, increased tolerance to uncertainty, and the removal of scepticism to basic forecasts or anticipations. Taking 'sequential' information and other realistic elements into account by delaying choice or action (which should not be confused with a strong preference for the status quo) seems quite sensible, even if they are not readily integrated in theoretical (especially mathematical) models.

A third area of neglect, where everyday notions of rationality conflict with popular specifications for consistent mathematical axioms of models, involves the acknowledgment of wider reflection as stressed in recent theories of regret. Basically, we might want to avoid comprehensive master plans in which pair-wise comparisons are assumed to be comprehensive and every possible contingency is noted, but which ignore the common phenomenon of recrimination. In particular, since choice is seldom isolated in time and individuals often reflect on the successes and failures of previous actions, we might then want to specify that 'when making a choice between two actions, the individual can foresee the various experiences of regret and rejoicing to which each action might lead, and that this foresight can influence the choice that the individual makes' (Sugden, 1986: 67). In everyday language, it is not irrational to recognize the strong likelihood of regret and rejoicing in 'second thoughts' or later reflection. This recognition translates into a call for the specification of different axioms to define rational choice in theoretical models or for a shift towards the alternative approaches associated with procedural rationality.

EQUIVALENTS

At the beginning of the present century, economists started to deal with the potential impacts of uncertainty by linking them to production, insurance, profits, trade cycles and other economic phenomena, and by trying to distinguish risk from uncertainty (Bigg, 1990). Most of us are now familiar with the influential contributions of Keynes (1921) and Knight (1921), and perhaps with the modest impacts of Edgeworth and Jevons on the discussion of probability by economists in this transitional period, but we are seldom aware of other interesting efforts (including those of Pigou, Lavington and Robertson). However, these efforts helped to set the scene for the later active exploration of uncertainty, certainty equivalents, and alternative portfolios of assets with only partially correlated risks in the 1930s and 1940s, which in turn came to significantly affect the content of economic theory up to the present time.

Early notions of a certainty equivalent were straightforward and did not necessarily involve the mean, mode or other measures of location for a probability distribution. Rather, they pointed to any acceptable means of summarizing a given group of contingent anticipations, a convenient device with which to analyse uncertain economic phenomena in a simplified manner. Keynes, for example, used the term 'expectation' for an entrepreneur within a simple operational frame:

> By his expectation of proceeds I mean ... that expectation of proceeds which if it were held with certainty, would lead to the same behaviour as does the bundle of vague and more various

possibilities which actually make up his state of expectation when he reaches his decision.

(Keynes, 1936: 24)

This treatment ignores the procedural aspects of forming estimates, neglects practical reasoning (as reflected in the consequences of disappointment with short-run expectations), ignores heterogeneity, and is clearly unsuitable for interpreting some important types of economic phenomena (Hart, 1941–1947). Critics such as Friedman insist that models involving certainty equivalents

> cannot explain the forces that lead corporations to make the kinds of decisions about how to finance themselves .... Again, there is no certainty-equivalent that can rationalize the existence of insurance companies; in a world of certainty, they would have no place. Uncertainty is of the very essence of phenomena like these.

(Friedman, 1949: 196)

Hart is even more scathing in his dismissal, in part because he stresses the sequential nature of information gathering and the benefits of flexibility.

> The certainty equivalent is a will-o'-the-wisp .... the business policy appropriate for a complex of uncertain anticipations is different from that appropriate for any possible set of certain equivalents. Trying to frame monetary theory in terms of certainty equivalents means leaving out specific reactions to uncertainty – which happen to be of fundamental importance for monetary theory. Furthermore, it leads to absurdities if pressed too far.

(Hart, 1947: 421)

A more sanguine attitude is expressed by Lange (1944: 29), who advocates that uncertain prospects be initially characterized by their most probable value, the mode, while any extreme values are disregarded. Earlier, Pigou, Hicks and Marschak had illustrated use of the mean for one-parameter representations of uncertainty. Lange considered the mode to be 'a more realistic descriptive device because an idea of it can be formed without any computation, by mere ranking' and 'it does not require the probabilities to be measurable'. In a similar vein, he recommends the 'practical range' of variation (instead of the variance or standard deviation) for two-parameter representations of uncertainty. The mode and some practical range of variation can be linked by a convenient trade-off or indifference relationship. Suppose a fixed value for U(M, R) identifies a group of alternative but equally-preferred prospects with differing modes M and practical ranges R. Then, a particular value of M corresponding to U(M, 0) is the certainty equivalent for the group and the difference [U(M, R) - U(M, 0)] is a 'risk premium' for the prospect with particular values of M and R.

While Lange's recognition of computational difficulties and po-tential truncation of probability distributions seems sensible (if rationality has a practical aspect), the use of only mode and practical range may seriously understate the information available to decision makers. Skewness is an obvious difficulty (Sinn, 1983: 48) for any myopic dependence on these two parameters (or on any other two parameters). Given this difficulty and those identified by Hart, Friedman and others with economic applications (and given the mathematical convenience of mean and variance), the adoption of certainty equivalents has fallen out of favour. Indeed, the *New Palgrave* (Freixas, 1990) reserves the term for a quite different notion, which we consider in the final section below. We return to the possibility of certainty equivalents in a wider assessment of expected values, partial ignorance, and probability representations when Shackle's kaleidic vision of choice and decision is discussed in the next chapter.

PORTFOLIOS

A partial identification of economic profits with uncertainty occurred more than seventy years ago (Knight, 1921). It was soon established that probabilistic elements are unavoidable in dynamic economic theories which avoid the excesses of perfect foresight. Three signifi-cant aspects of probability emerged (Hicks, 1931, 1934, 1935). First, financial assets (or income-bearing activities, or lines of business) seldom occur in isolation so the attendant risk for any economic choices (and activities) reflects the combined risk which is associated with a financial portfolio (or batch of activities). Second, because of partial independence of returns on financial assets (or individual sources of profitability), this composite risk involves the offsetting of diverse elements through a lack of correlation over time. Finally, there may exist a clear trade-off between the expected yield of the portfolio and the variability or volatility of this yield. The recognition of these three aspects launched a common framework for microeco-nomic portfolio analysis or mean-variance theory, in which cardinal-utility notions could also be integrated. Use of simple analogies also extended the framework to macroeconomic phenomena (Tobin, 1965; Feldstein, 1969) including liquidity preference and other features which are associated with money – 'the demand for money, being a stock demand, must proceed in terms of a balance-sheet theory' (Hicks (1982: 7), recalling his introduction of portfolio analysis to monetary theory about a half century earlier).

For financial securities, investors are typically presumed to have (form, or find in a costless way) probability distributions illuminating their potential future performance. Until recently, such distributions were assumed to have finite means and variances, and the individual investor's preferences in determining the levels and composition of their portfolios were presumed to be primarily affected by the size of such probability-based parameters. These preferences might also be qualified by constraints of the Friedman–Savage type, and by a criterion for 'rationality' of optimal choice which looks beyond simply expected value because 'diversification is both observed and sensible; a rule of behavior which does not imply the superiority of diversification must be rejected both as a hypothesis and as a maxim' (Markowitz, 1952: 78).

In practice, portfolio analysis for securities is identified with a search for the best combination of mean value (E or M) and dispersion or variance–covariance matrix V, where these parameters are either estimated or based on 'wise' judgment – with quadratic-programming techniques establishing how search might be limited to efficient and attainable (E, V) combinations.

> A portfolio is inefficient if it is possible to obtain higher expected (or average) return with no greater variability of return, or obtain greater certainty of return with no less average or expected return.
>
> (Markowitz, 1959: 129)

Clearly, the choice of particular parameters is somewhat arbitrary here and will usually be affected by a host of factors, including 'cost, convenience, and familiarity' as well as basic notions of what constitutes good or bad features of portfolios. Variance, for instance, may be inappropriate for non-symmetric distributions and suffers from the problem of giving the same attention to extreme high and low (perhaps negative) values.

Apart from the practical choice of representative parameters, any use of portfolio analysis requires the nature of probabilities and the stability of actual probabilistic beliefs to be clarified. A mature Hicks discounted the possibility of repetition when insisting:

> Investments are made, securities are bought and sold, on a judgement of [non-frequentist] probabilities. This is true of economic behaviour; it is also true of economic theory. The probabilities of 'state of the world' that are formally used by economists, as for instance in portfolio theory, cannot be interpreted in terms of random experiments. Probability in economics must mean something wider.
>
> (1979: 107)

Hicks noted the potential occurrence of situations for which available evidence is insufficient and some probabilities are non-comparable, especially the probabilities of extreme values, adversely affecting the calculation of expected values and variances. Markowitz (1959: 272) also stresses personal probabilities and recognized their lack of firmness – 'the existence of personal probabilities does not necessarily imply that as of the moment, the individual is positive that his beliefs are "good beliefs" ' and he may admit the presence of potential bias, hopefully eliminated *on average* over a lifetime of choices, but perhaps corrupting particular choices of portfolios.

Unfortunately, portfolio analysis has been identified with the crude inclusion of E and V in the utility function of the investor and with a powerful myopia which ignores other characteristics of relevant probability distributions. The young Hicks envisaged a wider framework, but quickly succumbed to the convenient EV approximation, as did many of his successors:

> The various frequency-curves of different investments could be arranged in order of preference . . . . In order to investigate the nature of this scale, the shape of each frequency curve may be studied by means of its *moments* – in the statistical sense. Each curve could be rigidly defined by taking a sufficiently large number of moments, and an approximation to the situation obtained by taking a small number.
>
> (1934: 195)

The long preoccupation with just two moments suggests a strong need for tractable mathematics and fashionability rather than awareness of the appropriate probability specifications. If normality is sufficiently widespread (perhaps for the logarithms of potential returns), then a focus on the first and second moments in portfolio analysis does little harm.

'Fatter tail' and the other stable but symmetric distributions (whose presence in financial markets has been discussed by Fama and Mandelbrot) can still be handled in a simple way by replacing the variance or standard error by other finite measures of dispersion. Asymmetries, bimodality and truncations, all possibly realistic features of the relevant probabilities, are much more difficult to handle – as are shortages of evidence, transactional costs, any temporal instabilities, special historical circumstances, and the learning failures which inhibit rational choice in practice. Hicks noted his own major concerns over the EV convention:

> many of the 'prospects' involved in actual business decisions are by no means normally distributed, but are highly skewed; and
> ... when we speak of risk, it is the skewness of the distribution, not the variance, that we principally have in mind.
>
> (1977: 166)

Similarly, advocating the use of a third parameter, Hicks concluded, in regard to the two-parameter model:

> one is left with an obstinate feeling that it does not go far enough. Why should the first and second moments of the probability distribution be all that we must take into account? It will make things harder if we try to go further; but there are some tiresome issues which we shall not clear up unless we attempt to do so.
>
> (1967: 117)

Surprisingly, little action has been taken on this suggestion in the thirty years since it was first expressed (although the likelihood of negatively skewed prospects for risky investments is clearly high and the possibility of normally distributed prospects is low), so any obstinate feeling on the need for more parameters may be specific to Hicks and not widely realized.

RATIONAL EXPECTATIONS

During the 1920s, the excesses of some prominent economists in providing the potential causes of cyclical fluctuations stimulated an important demonstration that such fluctuations can be simulated by combining simple linear operations with white-noise sequences (i.e.

61

collections of successive shocks or disturbances which are uncorrelated and have constant means). The artificial 'mimic curves' exhibit cyclical fluctuations which are similar to those found in the actual data for economic variables, but they are completely free from any contributions of economic theory (Morgan, 1990). The mimic curves offer a simple reminder that any causal analysis of data is fraught with technical difficulties and that it is much too easy to detect false 'structures' in temporal patterns.

The simple stochastic processes introduced by Yule (1926) and Slutsky (1937) – later popularized by Frisch (1933), Wold (1954) and Whittle (1963) – could also be used to demonstrate how routine manipulation of data (e.g. in the various forms of seasonal adjustment or smoothing, temporal aggregation, and elimination of trends) might introduce misleading false fluctuations such as 'long waves' (Dobb, 1939; Wald, 1939; Bird et al., 1965). Thus, looking back to the end of the 1960s when rational expectations first attracted any significant attention and when econometrics became a standard part of graduate training for economists, the subsequent assimilation and rapid promotion of such linear processes (and the associated conditional expectations and optimal predictors) as major ingredients of both theoretical and empirical macroeconomic models would have been very difficult to predict. However, by the mid-1990s, these linear processes (now firmly embedded in rational expectations, econometrics and fashionable discussions of unit roots and stochastic macroeconomic trends) were ubiquitous and any cautionary origins quite forgotten. Their popularity (expressed in a general specification of single or multiple linear difference equations, with constant coefficients and persistently disturbed by random shocks) usually comes with some awkward and ill-considered baggage – excessive normality, covariance-stationarity and (anti-historical) ergodicity, and myopic preoccupations with conditional expectations, but also a devious neglect of aggregation, lapses in information, and relevant time intervals.

The theoretical attractiveness of ergodicity (i.e. of probability-preserving transformations across time) for inferential models enjoying wide generality is clear in the general context where it is believed 'a scientific theory bound to an origin in time, and not freed from it by some special mathematical technique, is a theory in which there is no legitimate inference from the past to the future' (Wiener, 1949: 26). However, the cost of ergodicity is substantial, since ergodic models must inevitably lack topicality or historical conditioning – the ability to address any particular episodes and experiences which arise from particular disturbances (Hicks, 1967: 157) – and are insensitive to any evolutionary changes in economic variables and in the temporary relationships which usually connect them. The history of estimated macroeconomic equations over the last half century, including the instabilities of linkages apparently connecting annual price (wage) changes with unemployment, money demand with national output, or levels of government deficits with the restraint of entitlement programmes, give little encouragement to the common assumption that systemic ergodicity (rather than systemic instability or historical context) adequately characterizes the macroeconomic systems within which we live.

Interest in 'rational' expectations (and in related 'implicit' expectations) was launched by Muth (1961) and Mills (1957, 1962), drawing on their separate research involving stochastic models of industrial operations. Muth's classical paper is extraordinary in its attitude towards probabilistic specifications and rationality. Along the lines followed by Friedman and Savage in their empirical validation of the EU theory, Muth outlines an abstract theory and seeks to show that two broad qualitative implications are 'as a first approximation – consistent with the relevant data', while providing a clear statement of the specialized form which underlies a 'rational expectations hypothesis' (REH). On a descriptive basis, he suggests that 'expectations, since they are *informed* predictions of future events, are essentially the *same* as the predictions of the *relevant* economic theory' and presumes that 'dynamic economic models do not assume *enough* rationality' (Muth, 1961: 537; italics added). More precisely, he suggests that the 'expectations of firms (or, more generally, the subjective probability distribution of outcomes) tend to be distributed, for the same information set, about the prediction of the theory (or the "objective" probability distribution of outcomes)' while 'information is scarce, and the economic system generally does not waste it'. These bold assertions seem excessively optimistic as regards to the quality of economic theory and the use of information by firms.

For his specialized form (e.g. for price fluctuations in an isolated market), Muth assumes linear equations, all disturbances being normally distributed, and certainty equivalents – as described for linear decision rules in the final section below – exist for all predicted variables; i.e. the dynamic behaviour of economic variables is always characterized by linear processes of the simple type introduced to economists by Yule, Slutsky and Frisch, and all expectations are identified with conditional expectations (optimal predictors within the ergodic and covariance-stationary systems of disturbed difference equations, where all variables have ARMA or Wold representations). To avoid speculative excesses and insider trading, rationality is obtained by imposing the requirements that 'aggregate expectation of the firms is the same as the prediction of the theory' and that 'the expected price equals the equilibrium price'. As regards the realistic view of markets, this stochastic framework is on a par with any of Aesop's fables but, seemingly, it is still highly attractive to modern economists.

Muth advocates his extreme form of 'rationality' on three grounds: it is (in principle) applicable to all dynamic problems, even moderate non-rationality of inherent expectations, it will lead to speculation and the creation of information-selling firms (both now common), and it can be modified to deal with any systematic biases, incomplete or incorrect information, and poor memory. But the only real test was whether the 'theories involving rationality explain observed phenomena any better than alternative theories'. Since variables are explicitly taken as deviations from their equilibrium values, it is difficult to see how this ultimate standard may be applied in practice. However, one prominent supporter of the REH perspective (Sargent, 1983: vii) argues that, within twenty years, Muth's ideas were 'understood and accepted, and they have now come to dominate important areas of macroeconomics and dynamic econometrics'.

Surprisingly, any substantial impact on econometrics has come from the forceful attack on 'Keynesian macroeconometrics' (Lucas and Sargent, 1981) and, in particular, from the suggestion that the structural forms of all large macroeconomic models lack autonomy because of the presence of cross-equation restrictions which spread the severity of parametric instability and thus effectively reduce model capabilities for producing policy evaluations or forecasts. This attack has some merit as a counterbalance to past excesses in the use of such models, but a widespread parametric instability and the need to continually recompute many parameters are difficult to reconcile with any forward-looking promotion of the alternative stochastic framework, which is so totally devoted to ergodicity and stationarity! Certainly, the theoretical suggestions provided by Lucas (1987) do not seem to leave much scope for econometrics or for a persistence of the 'rational expectations revolution' as it affects dynamic and probabilistic elements beyond the narrowest confines of economic theory. But then it is difficult to see why his earlier suggestions (Lucas, 1972), where variables are defined as deviations from a 'natural rate' (and thus unmeasurable), had so much influence on academic opinion and stimulated attempts to find an operational econometric embodiment of the REH (or natural rate) perspective for macroeconomics.

Perhaps this development simply reveals an unfortunate distaste among many economists towards clarifying 'method' and establishing the appropriate empirical bases for their research efforts, beyond token references to broad qualitative findings. Any brief glimpse at the econometric studies re-printed in Lucas and Sargent (1981), seminar reports given in a special issue of the *Journal of Money, Credit and Banking* (November 1980), or the relevant articles in *Econometrica* and the *Federal Reserve Bank of Minneapolis Quarterly Review* over the last decade will quickly reveal how little we have achieved by combining real data with rational-expectations models. Not surprising, then, that calibration is currently being set up as an alternative to estimation for small macroeconomic models.

For theory too, some difficulties with rational expectations have attracted attention. First and foremost, the awkward neglect of aggregation was clearly visible from the start in the ease with which subjective probabilities were attached to markets. Not only do most aggregate time series fail to exhibit similar structures to those of constituent parts – e.g. adding two variables with simple but different AR processes may produce an aggregate process with ARMA characteristics – but a host of other complications occur in moving from an individual to aggregate outcomes. Some of these are dealt with by contributors to Frydman and Phelps (1983) and by Haltiwanger and Waldman (1989). Also some models involving rational expectations may have no solution or multiple solutions (Taylor, 1977; McCafferty and Driskill, 1980), especially if non-linearities are introduced. The models have considerable difficulty in assimilating heterogeneity, learning (Blume and Easley, 1982; Bray and Savin, 1986), uncoordinated expectations (Guesnerie, 1993), unstable equilibria (Carlson, 1987), asymmetric information, incomplete markets, discontinuities of demand in relation to price (Green, 1977), and other realistic features of the economic context (Taylor, 1985; McCallum, 1993).

INCREASING RISK

As the brief account of moral expectation and expected-utility theories in the last chapter made clear, many economists have often focused their attention on the potential occurrence and implication of diminishing marginal utility of income, wealth and money (more generally, on the presumed pattern for 'curvature' of individual utility functions). The irreversible shift to stochastic economic situations from their deterministic counterparts (and to common mathematical representations rather than less formal ones) produced a need for some simple means of summarizing this curvature, forms of stochastic preference and dominance, and of clarifying the notion of increasing risk – generally with the Friedman–Savage 'wiggly utility curve' and their treatment of risk aversion or the state-preference framework (Hirshleifer, 1965, 1966) providing a suitable initial point of reference.

In the mid-1960s, two indices of absolute and relative risk aversion, e.g. $r(Y)$ and $r^*(Y)$, were introduced by Arrow (1965) and Pratt (1964). Later extended – e.g. by Yaari (1969), Ross (1981), Machina (1982), and Machina and Neilson (1987) – these indices are defined in terms of the first and second derivatives of the utility function $U(Y)$ with respect to income or wealth Y:

$$r(Y) = - U''(Y)/U'(Y) \text{ and } r^*(Y) = Yr(Y).$$

They can be simply interpreted as measures of local risk aversion (risk aversion in the small), but also connected to corresponding notions of increasing or decreasing risk aversion when treated as increasing or decreasing functions of Y – measures to be used while preferences and utility functions do not change, momentary measures or ones appropriate to static conditions. Probability intrudes when the basic utility function is extended to uncertain prospects and preferences are summarized by a cardinal NM index.

Potential use of the measures is not restricted to preferences satisfying the common assumptions of expected-utility theories, but they suffer from many problems similar to those we identified with the two-parameter characterizations of portfolio analysis above, including referential difficulties with certainty. Basically, the concepts of increasing risk and attitudes to risk are not readily described by a few summary statistics. Even the simple question of whether one random variable is 'more variable' than another random variable (i.e. the comparison of two probability distributions) raises quite complex issues, as explored by Rothschild and Stiglitz (1970), while the transitivity and other properties of preferences in an uncertain environment need not be as strongly determined as expected-utility theories typically suggest.

An alternative and more complex approach to the comparison of distribution functions and the ordering of uncertain prospects was introduced by Hadar and Russell (1969), Hanoch and Levy (1969), and Rothschild and Stiglitz (1970) – and is conveniently summarized by Machina and Rothschild (1990: 232–7), who list some other useful references – through some technical devices, such as the focus on mean-preserving spreads, and new concepts of stochastic dominance and compensated increases in risk. Again, although newer approaches combine various intuitive notions of interest in regard to increasing risk, their theoretical results are still primarily concerned with comparative statics, and thus remain ill-suited to dynamic aspects of either preferences or probabilities – full of sound mathematical theorems, but seemingly remote from any extensive practical applications.

With stochastic environments, it becomes possible to assess general tendencies rather than inevitable consequences when dealing with the potential transitivity of preferences. Suppose we use the notation $p(X, Y)$ to represent the probability that X is preferred to Y by an individual economic agent. Thus, if $p(X, Y)$ is at least a half, then we can say that X is preferred more often than not in relation to Y (alternatively, 'is usually preferred to Y'), and the corresponding concepts of stochastic transitivity can be defined in terms of constraints on such probabilities. We could, for example, introduce an axiom of the type which requires that if both $p(X, Y)$ and $p(Y, Z)$ are at least a half, then $p(X, Z)$ also is at least a half – i.e. if X is usually preferred to Y and Y is also usually preferred to Z, then X is usually preferred to Z (weak stochastic transitivity). Strong and moderate forms of transitivity can be readily introduced by amending the implication to the probability of X being preferred to Z by the maximum, or minimum, of $p(X, Y)$ and $p(Y, Z)$ respectively. Development of such forms seems a valuable means of enlarging conventional axioms for rational choice towards normal or average tendencies rather than unwavering implication in uncertain situations – a sensible movement towards the modelling of less precision.

DECISIONS: RULES AND RATIONALITY

Operations research and management science began to appear in the 1950s, when techniques were sought to assist managers of firms in making better judgements, developing sound control and (internal and external) information systems, and re-assessing their marketing and production options. Some special problems with the control of inventories and the smoothing of production drew attention both to practical feedback relations and to the clarification of dynamic costs, in addition to attracting more attention to the goals and practices of decision-makers after a dismal period in which tedious debates on imperfect competition and the theoretical relevance of mark-ups or normal costs had been ineffective in removing business decisions from the opaque 'black box' of timeless microeconomic 'optimality', characterized by unqualified profit maximization and cost minimization.

With the new search for normative guidelines, there came a direct recognition of the inherent uncertainty affecting choices, actions and processes – and a strong stimulus to accept more computational and cognitive restrictions, to investigate useful approximations, and to re-phrase rational or sensible conduct in terms of heuristics. Two early features of this transformation of inquiry were a clarification of the limited scope for 'certainty equivalence' and the tentative promotion of linear decision rules, which came from optimization of convenient approximations to the industrial problems and from the need for practical computability. These decision rules are illustrated by Simon (1952), Theil (1957, 1964) and Holt *et al.* (1960), and their primary mathematical aspects are briefly surveyed by Freixas (1990). More generally, the transformation led to a better understanding of the alternative views of rationality, outside traditional confines of max-imization or minimization, and both the feasibility and limitations of the simple decision rules.

When dealing with portfolio theory above, we distinguished between a utility function expressed in terms of some parameters of a probability density function and the expected value or mean of a utility function defined in terms of the returns to assets held in a portfolio. These are components for a general issue: when can the arguments in a given objective function (utility, cost, profit) be replaced by parameters from an associated probability density function so that optimal values of the modified function are still indicative of values for which the expected value of the initial objective function is optimized? Specifically, suppose $O(X, Z)$ is a relevant objective function and the individual agent is assumed to maximize its expected value $E[O(X, Z)]$ for conditional values of Z, subject to any restrictions described by setting $r(X, Z)$ equal to zero. Suppose $Q(X, Z)$ represents the corresponding function with conditional probability parameters X substituted for X. Then the Simon–Theil results from the 1950s provide a simple linkage between the optimal values associated with the two objective functions in special cases, which they specify.

Currently, the theoretical certainty-equivalent problem is concerned with the potential 'identification of the conditions under which ... an isomorphism between the optimal decisions under uncertainty and the optimal decisions in an equivalent certainty context hold' (Freixas). There are few difficulties for either static or dynamic models if objective functions are quadratic and supplemental restrictions are linear, but these are very severe constraints on the range for theoretical applications of these equivalents. A small extension is given by Malinvaud (1969) and Theil (1957) to 'first order' equivalence but there seems little basis for presuming much more scope for simple isomorphisms. Fortunately, there is an alternative procedural justification for an interest in decision rules – a strong justification which was clearly exhibited throughout the early influential efforts of Holt, Modigliani and Simon (who stressed practical applications of dynamic programming, feedback mechanisms, and computable algorithms rather than purely theoretical ones). Adaptive rules for adjusting inventories, production and employment were initially developed as approximations to be tested by applications at the plant-level, and their relative success was not to be judged by theoretical elegance but rather by simple convenience, feasibility and effectiveness in concrete situations (Simon, 1976). There was no illusion here of substantive optimality. Instead, practical success was associated with the concept of satisficing and computability, and ultimately with a bounded (perhaps descriptive) rationality which was somewhat distant from the stress of deductive reasoning based on consistent collections of axioms.

This shift away from substantive rationality and towards more realistic (but still simplified) characterizations of reasoning and decision-making is extensively described in Simon (1982, 1983) and Egidi and Marris (1992). Its influence in promoting a joining of psychological and economic research (a 'cognitive revolution') is strong, as should be clearly visible to most observers from the tone and content of major assaults on expected-utility theories. The term 'bounded rationality' was introduced by Simon to

> focus attention upon the discrepancy between the perfect human rationality that is assumed in classical and neoclassical economic theory and the reality of human behaviour as it is observed in economic life. The point was not that people are consciously and

69

deliberately irrational, although they sometimes are, but that neither their knowledge nor their powers of calculation allow them to achieve the high level of optimal adaptation of means to ends that is posited in economics.

(Egidi and Marris, 1992: 3)

In part, this shift to procedural and bounded rationality reflects a firm movement of qualifications on model implications (auxiliary empirical assumptions which permit these implications to deal with real economic 'complications') from post-theory adjustments to a sensible within-theory recognition (ibid.: 4), and thus to amend chronic assumptions of general omniscience.

There is a fundamental issue here of whether optimal decisions are reasonable (Einhorn and Hogarth, 1981: 55). The emergence of bounded rationality theories has established that, because of the overwhelming complexity of uncertain economic environments and the conditional nature of any theoretical 'optimality', there is an important methodological choice between pursuing optimal models (whose direct implications must be subsequently qualified for real applications) or developing heuristic models with more realistic characteristics. The key features of this choice are a recognition that conditionality always significantly affects the promotion of any normative optimality, and an acceptance of sufficient room for a plurality of modelling procedures to be followed – 'there is no single sovereign principle for deductive prediction' (Simon, 1986: S223).

Because of the hazards of irresponsible gerrymandering with convenient auxiliary assumptions, the testing or confrontation of many theoretical models (especially neoclassical models) with real evidence is flawed – recall the modest use of evidence in Muth's initial account of rational expectations and the Friedman–Savage tests with stylized facts for their specification of curvature for individual utility functions. Procedural rationality, at least, begins with an inspection of major 'factual' aspects of economic situations – including the nature of human reasoning (cognitive architecture), transactional and computational costs, and some intelligent adaptive processes.

4

IGNORANCE, SURPRISE
AND VAGUENESS

Many economists still accept the assertions of Marschak (1954) that 'uncertain knowledge is not ignorance' and 'empirical social science consists of statements about probability distributions'. The potential invalidity of such strong assertions is at the heart of the wide range of radical views reported in the four sections below. Our discussion considers views which have been associated with major challenges to the primacy of conventional probabilistic-based perspectives, giving more attention to the recognition of vagueness, imprecision, imagination and other aspects of mental activity, contextual issues (including evolving goals, particular circumstances and information), and the troublesome impracticality of applying probability theories and some experimental evidence to major decisions and risk evaluations covering substantial hazards.

In the first section, we look at Shackle's idiosyncratic view of decisions, in which frequentist probabilities are discarded for a better appreciation of epistemic states, the use of imagination to clarify (and, perhaps, to produce) the options and consequences of current economic decisions, and the potential uniqueness of many important ('one-shot') decisions. This view evolved, in part, from Hayek's treatment of personal knowledge, economic subjectivism as applied to expectations (Lachman, 1990) and other unconventional themes involving uncertainty which emerged in the 1930s and 1940s, but also from the willingness of Shackle to creatively address the limitations of human attention, the role of surprise, awareness of best and worse sequels, and the realistic pressures of timeliness. For Shackle, any sensible modelling of actual decisions and decision-making requires the explicit recognition of personal mental activities, especially imagination, and the accommodation of 'unknowledge' rather than optimal choice among a fixed collection of predetermined options:

> Decision ... is the imagining of rival paths of affairs; the assigning of these paths to the respective actions, amongst those the decision maker can envisage, which seem to make them possible; and the resolving upon that action which, at best, offers a sequel more powerfully out-weighing what it threatens at worst, than any rival action. We thus suppose the decision maker necessarily to originate the entities, each an action and its skein of rival imagined sequels deemed possible, amongst which he will choose. The business of inventing such entities will be stopped only by his reaching a deadline for decision.
>
> (Shackle, 1986a: 58)

Subjectivism is evident here in the attempt to elucidate any socio-economic phenomena in terms of their inherent meaning to individual economic agents, differing across such agents and firmly embedded in their mental activities.

In the second section, we turn to the development of fuzzy set theory, fuzzy logic and fuzzy concepts to deal with inexactness, giving special attention to the prominent and persistent role of Zadeh in the corresponding assault on the value of both classical logic and prob-ability theory as primary components for imprecise theoretical analyses (also including their applications to expert systems and artificial intelligence). Zadeh's initial concern was simply to clarify the potential of unconventional theories of fuzzy mathematics 'for dealing with phenomena that are vague, imprecise, or too complex or too ill-defined to be susceptible of analysis by conventional mathe-matics' (prologue to Kandel, 1986: viii), given his firm impression that 'much of human reasoning is imprecise in nature, and most of it is not amenable to formalizations within the framework of classical logic and probability theory' (ibid., vii). He also calls into question the expressiveness of probability as a convenient language for uncertainty, by offering an alternative approach based on the new and evolving concepts to be found in the theory of fuzzy sets, fuzzy probability and fuzzy syllogism. In regard to logic, he insists that there exists an 'excessively wide gap between the precision of classical logic and the imprecision of the real world' (Zadeh, 1984 [1987: 19]).

Zadeh holds that the conventional quantitative techniques are intrinsically unsuited for dealing with complex humanistic systems, unlikely to have relevance for 'real-world societal, political, economic,

and other types of problems which involve humans either as individuals or in groups' (Zadeh, 1973: 30). This view reflects a distinctive
image of human reasoning which focuses on labels for fuzzy sets rather
than numbers, prefers approximation of data to precision, and severely
restricts attention to specific aspects of task-relevancy. His theoretical
preoccupations with fuzzy logic and fuzzy mathematics, primarily,
seek to imitate human tolerance of imprecision in a wide range of
potential applications – e.g. artificial intelligence, linguistics, many
human decision processes, pattern recognition, psychology, medical
diagnosis, information retrieval, law, economics and other social
sciences. The common feature of these particular areas is their
complexity as affecting the inherent simplicity or imprecision of
basic concepts. Zadeh, for illustration, points to the basic fuzziness
which characterizes the concepts of:

> recession and utility in economics; schizophrenia and arthritis in
> medicine; stability and adaptivity in system theory; sparseness
> and stiffness in numerical analysis; grammaticality and meaning
> in linguistics; performance measurement and correctness in
> computer science; truth and causality in philosophy; intelli-
> gence and creativity in psychology; and obscenity and insanity
> in law.
>
> (1976a: 249)

In the third section, we focus attention on possibilities and approximate reasoning from three perspectives. One view, due to Shackle,
stresses the prominent role of imagination in creating and delineating
judgemental possibilities rather than probabilities and (perhaps)
understates the influence of experience and the incidence of collective
decision-making. A second view assembles mathematical theories of
possibilities, partially stimulated by Zadeh (1978) and the rapid
extension of fuzzy concepts to inexact mathematical and statistical
applications as illustrated, for example, by various contributions to
Yager (1982). The final view reflects a more 'naturalistic' or pragmatic
approach to actual decision-making, which is described by Klein *et al.*
(1993). This approach stresses performance in the dynamic context of
evolving situational factors, non-discrete choice and embedding of
decisions in broader goals or 'task cycles':

> much effort is devoted to situation assessment, or figuring out
> the nature of the problem; single options are evaluated

sequentially through mental simulation of outcomes; and options are accepted if . . . satisfactory.
(Orasanu and Connolly in Klein *et al.*, 1993: 6)

In contrast to the perspective of Shackle, experience has greater significance in the naturalistic approach and most decisions are not considered to be primarily one-shot efforts. Instead, they are viewed as normally part of an ongoing process involving sequential learning through experience and effective responsiveness to novel facts and goals.

In the final section, we point to the challenges of hazardous situations to the practical notions of acceptable or unacceptable risk. Basically, our concern involves the potential feasibility and value of applying conventional probabilistic techniques in risk evaluation and environmental epidemiology by government agencies, professional associations, public-interest groups, and large firms – e.g. in connection with smoking, AIDS, nuclear power, radon emissions, pesticides and chemical or carcinogen exposure, low-frequency electromagnetic fields, genetic engineering, and the storage and treatment of hazardous waste. Interesting illustrations of the normal forms of discussion of risk evaluation and management in this context are provided by Fischhoff *et al.* (1982), Wakeford *et al.* (1989), Kaser (1989), Hodges (1987), Baron (1993), Lane (1989), and contributions to a special issue of *Statistical Science* (August 1988). Again, as noted in an earlier chapter, the non-technical or everyday language of risk (especially catastrophic risk) and uncertainty may seem markedly distant from counterparts found in the fashionable academic interests and principal themes of many professional journals.

Our brief treatment of ignorance, imprecision and vagueness is inevitably incomplete and limited in scope. Some relevant topics are left to be raised in Chapter 6. These include unreliable and indeterminate probabilities (Gardenfors and Sahlin, 1982, 1983; Levi, 1974, 1979), knowledge in flux and epistemic entrenchment (Gardenfors, 1988), belief functions (Shafer, 1976a, b), the vagueness represented by the difference between upper and lower probabilities (Koopman, 1940a; Good, 1962a), and incomplete probabilities (Hicks, 1979; Keynes, 1921). We also omit more discussion of the vagueness or ignorance which is recognized by the popular reliance on either uninformative priors in Bayesian statistics or the principle of indifference and insufficient reason. A more comprehensive account of relevant technical issues in regard to imprecise probabilities, uncertainty and ignorance is provided in the valuable treatment of Walley (1991).

All of the four principal areas which are addressed in this chapter – Shackle's kaleidic vision and his stress of imagination, Zadeh's fuzzy mathematics and linguistic ingredients, naturalistic decision-making in evolving or transforming circumstances, and the partial evaluation of hazardous risks – rely heavily on realistic attachments to justify their constituent elements. Shackle, for example, is preoccupied with creative but limited mental activity, the uniqueness, immediacy and unpredictability of many decisions, the self-destructive aspect of these decisions (which irretrievably alter economic situations), and thus he opts for additional realism rather than the audacious abstractions of static equilibrium theory (Shackle, 1958: 92–3) and comfortable conventions of probability theory. Similarly, Zadeh supports his interest in fuzziness by insisting:

> The ability of the human mind to reason in fuzzy terms is actually a great advantage. Even though a tremendous amount of information is presented to the human senses in a given situation … somehow the mind has the ability to discard most of this information and to concentrate only on the information that is task relevant.
>
> (1984 [1987:20])

However, the recent success of fuzzy mathematics and corresponding technologies owes more to their commercial applications, especially found with control systems and pattern recognition, in such diverse areas as washing machines, camcorders, air conditioners, computers, vacuum cleaners, navigational equipment, transit control systems, and car transmissions (Bezdek, 1993: 3). A comprehensive list of Japanese applications and some potential extensions are provided by Terano et al. (1992). We have already pointed to the realistic domains of naturalistic decision-making and risk evaluation, which are inevitably compelled to deal with the main features of human experience and real environmental aspects in their case studies.

IMAGINATION, INFREQUENT EVENTS AND SURPRISE

In normal economic situations, it seems clear that we know the probabilities of very few things and that we lack simple procedures for combining fragmentary knowledge to infer any precise numerical values for probabilities, whether personal or frequentist in form and whether coherent or not. Human computational skills, in any case, are bounded, biased and somewhat confused, while individual deliberation and consensus-seeking for group decisions are usually subject to significant time constraints. Further, many decisions that seem to involve numerical probabilities are essentially non-repetitive and thus yield few real opportunities for generating any crisp estimates of these probabilities, which may not be stable in any case. This awkward scenario contrasts with the sharp precision typically assumed in most theoretical models of stochastic choice or decision-making – where the goals, constraints, data, relevant information, and basic relationships are relatively unflawed by vagueness and where the economic agents' actions are assumed to be situationally rational.

The obvious implausibility of routine elucidation of numerical probabilities in unfamiliar and vague circumstances must weaken the actual ability of theoretical models to cast any substantial light on particular choices, decisions and actions. When we also invoke subjectivism as a basic research programme in social sciences, the environmental imprecision is augmented by spontaneous and adaptive mental autonomy – by transitory 'human purpose' or conjecture, and the subjectivity of data – so the feasible domain of probability for human conduct is drastically reduced, at best limited to groups of events and quite remote from case probability (Lachman, 1976: 57). For Shackle, the recognition of this restless 'human factor' yields a kaleidic process, not part of any equilibrating mechanism (in contrast to the evolving competitive pressures and discovery envisaged by Hayek and other economic subjectivists).

From a personalist perspective, there might still be *evolving* or *sliding* probabilities 'that are currently judged and ... can change in the light of thinking only, without change of empirical evidence' (Good, 1971: 124), but these differ markedly from the probabilities which are normally considered in theoretical models. For frequentists who stress repetition and pooled experience, the essential complexity of the kaleidic process described by Shackle provides just qualitative uncertainty and permits no meaningful role for the familiar concepts of 'population' and mathematical expectation. Instead of the knowledge represented by probability as relative frequency, Shackle points to the 'unknowledge' of a decision-maker, who is conscious that

even the foreground of his field of action in time-to-come is a shifting mist and that all beyond it melts into a void. He needs to assess, to express in some way his hopes and fears, what he can envisage at best and what he must be ready for at worst. Choice ... is not an experiment which can be repeated indefinitely under broadly unchanging conditions.

(Shackle, 1979a)

A past record of non-uniform choices and decisions is insufficient for establishing future chances. Indeed, repetition may be logically impossible because situations are irretrievably altered by actions, while seeking guidance from the outcome of a single trial has very questionable propriety if most decision-requiring situations occur rarely in any lifetime.

To most business men it falls only once or twice, or a handful of times, to have to decide upon the purpose, type, scale and location of an individual plant; most professional men choose a career only once; and so on. These occasions of choosing are spread at such long and irregular intervals that they cannot be treated as together forming a seriable experiment.

(Shackle, 1958: 37)

Kaleidic expectations can change comprehensively and radically due to experience of slight shocks or to new evidence, and kaleidic economics requires the imagining and study of potential adjustments stemming from momentary expectations and beliefs. The calculation of numerical mathematical expectations, on the other hand, must involve an acceptance of the confident assertion that contingencies are realized as divisible outcomes. This assertion is incompatible with the non-divisible, non-seriable and non-uniform context of any kaleidic process with precarious and ephemeral states and mutually-exclusive hypotheses.

What is the sense of a weighted average which adds together a hundred falsehoods and one truth after multiplying each by some irrelevant number?

(Shackle, 1958: 43)

Even with a less anarchic vision of probabilistic inference in complex situations, the obvious potential for unsettling discovery in mental reasoning reveals a fundamental weakness of any excessive reliance on mean or mathematical expectation. As noted by Glymour, for example, looking at a common phenomenon affecting human experience with discovery and learning:

one realizes that there is a new hypothesis that has not previously been considered. What rational humans do in that circumstance is change the algebra on which their degrees of belief are defined, and sometimes merely considering a new hypothesis causes substantial changes in the degree of belief assigned to other hypotheses.

(Glymour, 1989: 87)

Shackle rejects completeness of any list of hypotheses or imagined sequels for kaleidic processes, preferring to assume open-endedness and to abandon probability for alternative new concepts such as potential surprise, epistemic interval, disbelief and ascendency which can be identified with possibility rather than probability. The particular concepts do not readily facilitate mathematical or econometric analysis, and thus have often been excluded from fashionable theories. However, they partially capture important aspects of realistic business decisions in the presence of systemic uncertainty or essential complexity.

The imprecise domain or population of any agent's feasible actions, choices or decisions is of primary concern in Shackle's rejection of probability. Deliberation (affected by a host of relevant factors, e.g. limited memory, computational capacity, emotional and sociological stability) provides a means of both reducing and revising alternatives so that knowledge can be viewed as expanded by the process of choosing. Both Shackle and Levi (1972, 1986) stress the need to emphasize what an agent will choose *from*, which is not fixed until the final moment of choice according to Shackle because of the scope for creative imagination and the renouncement of some options. This final moment requires a psychic shift from an initial state with a group of hypotheses predicting the implementation of various options to a new state in which only one hypothesis is countenanced. Given the ephemeral nature of hypotheses/options, there is no obvious way of giving a probabilistic account of this deliberative activity or its eventual resolution. Further, because of the subjectivity of data used in this activity, decisions are neither predictable nor subject to subsequent appraisal by external observers as to their apparent rationality (Shackle, 1958: 20–8).

This kaleidic vision is unsettling for the two generations of economists who were trained to believe in the general applicability of probability distributions and who became committed to the comfortable Marschak assertions with which we began this chapter.

Substitution of the potential-surprise function for more familiar probabilistic alternatives and the partial abandonment of standard calculations for vague or disorderly kaleidic notions must lead to substantial discomfort, perhaps still avoided by 'as if' rationalization and the continual avoidance of those methodical issues that are associated with the status of realistic elements in promoting or validating economic theories and their predictions.

Besides Shackle's persistent attempts to clarify and extend this radical perspective, there are interesting descriptions and modified versions provided by Ozga (1965, ch. 5), Ford (1983, 1987, 1989), and the commentaries reported by Carter and Ford (1972) and Carter *et al.* (1957). The Carter–Egerton theory, for example, seeks to give a sensible description of the uniqueness, irreversibility and simplification of business decisions. Uniqueness stems from an identification with 'a particular organisation at a point of time, with a pattern of assets or available factors, of technology and of market opportunities which is most unlikely to be repeated anywhere else in the economic system' (Carter, 1972: 30). Simplification takes the form of omitting abhorrent and extraordinary actions, omitting those actions similar to others, and separating decisions into distinct stages – the domain of any contemplated outcomes is markedly reduced by ignoring unlikely circumstances or those with little relevance, difficult or unpleasant possibilities, and minor variations (ibid., 41–2) – and decision-makers can wait for some uncertainties and tensions to be resolved. Neglect of prospects with small probabilities has also been reported by Allais in his classical attack on neo-Bernoullian models.

INEXACTNESS AND FUZZY CONCEPTS

The active interest in fuzzy sets as an appropriate means of characterizing one notion of inexactness began about thirty years ago, when Zadeh (1962) first called for the development of a new mathematical apparatus involving 'fuzzy or cloudy quantities which are not describable in terms of probability distributions' and when Zadeh (1965, 1968), Goguen (1967, 1969) and others launched the search for meaningful analyses with inexact concepts. The success of subsequent efforts can be gauged from the contents of relevant textbooks – e.g. Kandel (1986), Terano *et al.* (1992), Zimmermann (1991), and others listed by Bezdek (1993) – and from contributions to the three major journals that were established to present research findings in this context, *Fuzzy Sets and Systems*, *IEEE Transactions on Fuzzy Systems* and

International Journal of Approximate Reasoning. Bezdek (1993: 5) also indicates the rapid world-wide growth of both conferences and professional institutions which focus more attention on fuzzy systems. Some of Zadeh's own prominent papers have been reprinted in Yager *et al.* (1987).

Fuzzy set theory has a close affinity to multi-valued logic as developed, for example, in the 1920s by Lukasiewiecz (Giles, 1976). However, the new theory has much wider options for its quantifiers – e.g. most, much, many, not many, very many, not too many, several, few, quite a few, large number, small number, close to, approximately, simple, serious, important, significant, accurate, and quite substantial. Mathematically, the fuzzy set is any class of 'objects' with a continuum of grades of membership. The set is characterized by a 'membership function' which attaches some non-negative real number to each constituent object – the closer this number is to unity, the higher will be the 'grade' of membership for the object in the fuzzy set. Zadeh (1965) defined the union and intersection of such sets, showed that the semantic and algebraic operations can be provided for them, and established a simple separation theorem. He followed this, three years later, by introducing the notion of a fuzzy event to which a numerical probability might be attached – the probability of the event is defined as the expected value of its membership function (Kandel, 1986: sect. 4.3; Zimmermann, 1991: sect. 8.2; and Terano *et al.*, 1992: sect. 6.1) and, therefore, not normally unique – hoping to 'significantly enlarge the domain of applicability of probability theory, especially in those fields in which fuzziness is a pervasive phenomenon' (Zadeh, 1968: 421). Other justifications for the probability of a fuzzy event have since been offered, for example, by Smets (1982a, b) and its formal representation continues to attract much attention.

Ultimately, Zadeh retreated from the conventional precision of most probabilities, preferring the term 'linguistic probabilities' for the objects of his interest, considering 'linguistic variables' instead of their numerical counterparts, and actively resisting the prevalent attitude whereby normal science must be both precise and quantitative:

> There are many things that cannot be expressed in numbers, for example, probabilities that have to be expressed as 'very likely', or 'unlikely'. Such linguistic probabilities may be viewed as fuzzy characterizations of conventional numeric probabilities.

And so in that sense fuzzy logic represents a retreat. It represents a retreat from standards of precision that are unrealistic.

(Zadeh, 1984 [1987: 25])

Instead of earlier conventions, the new use of linguistic variables (Zadeh, 1975a, b; 1976b), fuzzy conditional statements and several categories of fuzzy algorithms or ordered sequences of instructions (Zadeh, 1973) with intuitively-plausible semantic descriptions of the objects in fuzzy sets, came to provide approximate descriptions of objective functions, constraints, system performance, particular strategies, and other model ingredients.

This radical transformation of research methods convincingly challenged the premise that imprecision can be simply equated with randomness. Fuzziness cannot be seen as an immodest surrogate for probability endowed with greater flexibility or elasticity to cover various deficiencies associated with incompleteness or imprecision. Also fuzzy models do not replace probabilistic ones, which might be viewed as some special cases with the crispness found in two-value membership functions (Bezdek, 1993; Ralescu and Ralescu, 1984). Often economic environments will contain elements of both fuzziness and randomness – the main issue for research is to determine what methodological structure works best in specific circumstances, and the optimal choice may involve both fuzziness and probability, or perhaps neither of them!

Within frequentist theories of probability involving samples from a given population, ignorance can be envisaged as not knowing which member of the population is being sampled. On the other hand, with fuzzy sets and linguistic probabilities, the boundary of the population itself remains vague while inherent ignorance stems from such vagueness or imprecision as to what objects (individuals) are contained in the population. Clearly, the presence of one form of ignorance does not preclude the presence of the other.

POSSIBILITIES: IMAGINATIVE, THEORETICAL AND PRACTICAL

For more than fifty years, Shackle has insisted that economic decision-makers are primarily concerned with possibility rather than probability, holding that a notion of epistemic possibility is indispensable for describing imaginative choice as an exposure to possibilities and the exclusion of plural options. Specifically, he suggests that:

> Possibility is a judgement. It is a thought, but it is not entire knowledge .... Possibility is relevant to choice when some imagined evolutions of affairs are judged to be *not* possible as sequels to some choosable course of action .... Choice is *effective* when it can confer or withhold possibility.
>
> (Italics in original, Shackle, 1979b: 35)

Possibility is then identified with the absence of any discernible 'fatal' objections to acceptance within the current knowledge of the chooser. In trying to represent possibility as an analytical variable rather than simply a categorical one, Shackle linked it (inversely) to concepts of disbelief and potential surprise within an epistemic interval.

Such concepts seek to indicate how relatively surprised a decision-maker *imagines* he would be if some possible outcome were actually to occur. Zero surprise or disbelief is consistent with a perfect possibility, and 'maximum' surprise (set at some arbitrary value) or absolute disbelief indicates non-possibility and complete exclusion. The range between these two extremes for potential surprise is the 'epistemic interval' delineating possible sequels, and formal accounts of active choice might be developed by treating surprise as a continuous 'uncertainty variable' in that interval – with some further representations of relevance or ascendancy being needed to combine potential surprise and additional aspects of reasoning. Potential surprise acts as a non-additive indicator of relative decisiveness for the acceptance or rejection of hypotheses (Levi, 1972: 234).

The positive role of experience in affecting possibilities rather than probabilities is clear. Shackle insists that experience only suggests what can come to pass.

> If experience could tell us what will come to pass, we would be in a world of determinate history, choice-denying and choice-abolishing.
>
> (1979b: 59)

Given the important role of imagination in producing new prospects, the activity of choice may bring into being, and make effective, something that is not implicit in antecedents, and thus realistic analysis is drawn away from the conventional notion of externally-given, 'ready-made', collections of possible sequels to actions. Expertise, now associated with the flexibility to address changing situations and produce imaginative options or new possibilities, can then be clearly separated from experience.

The constructive elements of Shackle's formal framework (not an essential abstraction or a major concession to formalism) for describing choice among variable possibilities seem unsatisfactory, unable to sustain interest as a means of interpreting uncertainty and awkward to use – despite the descriptive attractions of his perspective for exploring the deliberative nature of human choices and decisions. The technical intractability of Shackle's graphical structures is a major factor in accounting for the rapid decline of attention given to his views, from the excitement displayed at the 1953 symposium reported by Carter et al. (1957) to the widespread neglect found today, and for the absence of adequate refinements over the half century that has elapsed since his kaleidic vision was introduced to economists in the Economic Journal (September 1939).

Potential surprise can be sketched as a flat-bottomed U-curve, functionally defined for a multi-outcome prospect over an effective range from substantial losses to substantial gains (Shackle, 1949: 13; Stephen, 1986: 47) with some 'neutral' outcomes corresponding to the central (unsurprising) flat portion. This graph contains both surprise and desiredness. To these two elements, Shackle adds the notion of ascendancy which reflects how the various gains and losses for various outcomes attract the attention of the decision-maker and which is formalized by a collection of ascendancy curves, zero for the central portion and rising with both gains and losses (Shackle, 1949: 23; Stephen, 1986: 48) as illustrated in Figures 4.1 and 4.2. The single potential-surprise function for any prospect is established by introspection. Its form matters because points of tangency with ascendancy curves are considered to be specially significant due to the inherent simplifications in choice which lead to such points being treated as primary foci for attention.

For any PS function, tangency indicates one primary focus for gain and another for loss, which in turn point to two standardized focus points (SL and SS) located on the horizontal axis of zero potential surprise by intersections with the particular ascendancy curves that are identified by tangency with the PS function. The graphical structure is completed by plotting gambler indifference curves across pairs of standardized focus gains and losses for different prospects (Shackle, 1949: 30: Stephen, 1986: 50), truncated at a boundary of focus losses beyond which levels are unacceptable. This cumbersome geometry is simply an expository device that does not eliminate inherent vagueness or subjectivity, for the calibration of both ascendancy and potential surprise within any epistemic interval is both 'unattainable

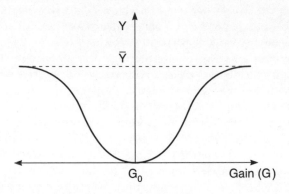

Figure 4.1 Potential surprise represented as a flat-bottomed U-curve

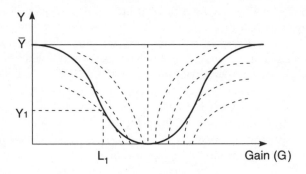

Figure 4.2 Tangency indicating primary focus for gain and loss

and unnecessary' (Shackle, 1986: 61). Criticisms of the framework are reported in Stephen (1986) and Carter *et al.* (1957).

Formal structures play a much more important role in Zadeh's approach to possibility theory, which also gives greater attention to communication and the fuzziness of language than is exhibited by Shackle. Thus, approximate reasoning as partially characterized by possibilities is markedly different for the two alternative schools of imprecise thought which are associated with potential surprise and fuzzy concepts. They share major objections to the ubiquity of probability for complex (human) situations, but reveal two distinct responses to their dissatisfaction with earlier conventions. Both offer firm realistic justifications for their substantial interest in the significance of imprecision, but the rapid development of fuzzy mathematics was more effective (and more actively sought) than any formalization of potential surprise and related concepts beyond the initial geometry offered by Shackle. Another difference between these two schools is found in the relative narrowness of contributions. Shackle dominates the treatment of possibilities from a kaleidic perspective, which has persistently remained close to his original conception, while Zadeh shares attention with a much wider group of important contributors in formal elaborations of fuzzy possibility and in extension of substantial applications to regression, discrimination, mathematical programming and other technical areas, for example. The fuzzy-set perspective continues to be transformed and re-assessed in comparison with other radical or 'non-standard' approaches to uncertainty and non-additive views of probability (Dubois and Prade, 1988a, b; 1993), quite different from the peripheral status of kaleidic theories.

The recent shift to fuzzy possibility from the conventional concepts of probability reflects a widening acceptance of Zadeh's assertion that the latter offer inadequate expressiveness for the real uncertainty of complex human systems, being flawed by the reliance on classical two-valued logic and an excessive presumption of quantification:

> most probabilities . . . are not known with sufficient precision to be representable as real numbers or, more generally, as second-order probabilities . . . . Although the theory of subjective probabilities does provide methods for elucidation of . . . numerical probabilities, it does not answer the question of how such probabilities are arrived at in the first place, nor does it come to grips with the issue of representation of imprecisely known probabilities as fuzzy rather than second-order prob-abilities.
>
> (Zadeh, 1986: 104)

Obviously, criticisms of this type have been resisted, e.g. by Lindley (1982a), but they seem to be receiving further support as interest grows in the formal treatment of fuzzy propositions, especially in connection with the active search for computational algorithms (both representational and inferential).

For successful implementation of fuzzy possibility, rules are required to represent the approximate reasoning and commonsense knowledge, linking uncertain 'events' to uncertain dependencies, that are supposed to be accommodated here – otherwise we are left with the apparent barrenness of the kaleidic vision. As indicated by the wide-ranging accounts of Kandel (1986), Terano *et al.* (1992) and Zimmermann (1991), formal structures for appropriate operations involving possibility theory are now established, with the measure theories of Sugeno and others providing a common basis for both possibility and probability since the mid-1970s, and with parallel concepts providing for more effective competition in many prominent areas of statistical analysis and operations research.

Ambiguous human thinking and subjectivity are dealt with by relying on fuzzy sets and membership functions, but the presence of vague boundaries must not imply that the practical logic is thereby tainted and vague. Standard rules and general definitions permit clarification and use of fuzzy differentiation, fuzzy integrals, fuzzy entropy, extrema for fuzzy functions, fuzzy control, fuzzy quantification, and fuzzy statistics – and analogous axioms can be found for possibility theory (Dubois, 1988). All that was needed to initiate these formal developments were sensible proposals for a mathematization of the intuitive concept of possibility and some clear demonstrations of their relative effectiveness in modelling particular circumstances where quantitative and qualitative forms of probability have been unsuccessful. Zadeh's association of a possibility distribution with fuzzy restrictions (i.e. elastic constraints on values assigned to a variable) and his assumption that a proposition in natural language could be interpreted as an assignment of a fuzzy set to some possibility distribution provided a convenient context for the developments.

The third approach to possibilities focuses neither on mental states nor on the mathematics of fuzzy propositions. Instead, it deals with realistic descriptions of the dynamic behaviour which might occur during actual decision-making in an evolving natural context and, hence, is termed 'naturalistic' by its proponents who seek better representations of behaviour relative to pragmatic criteria for effectiveness. As such, most accounts of naturalistic decision-making are anecdotal, stressing situational elements and the adaptive learning of possible actions, outcomes and scenarios. There is a strong kaleidic element in the perception that decisions occur in an ever-changing environment with many significant aspects being only partially known to participants. However, there is also greater stability than envisaged by Shackle due to some limitations on imagination and more sensitivity of economic agents due to the need for integration of wider issues in decision processes.

Similarity with the kaleidic and fuzzy perspectives includes a recognition of awkward elements which frequently complicate major choices and decisions. For naturalistic descriptions, these can involve ill-structured problems, uncertain dynamic environments, shifting and competing goals (often ill-defined), feedback loops, stressful time constraints, high stakes, multiple players, and organizational goals or norms (Orasanu and Connolly in Klein *et al.*, 1993: 7) – all of which affect the delineation, awareness, emergence and stability of possibilities. Each episode is unique and extended over time, but sufficient expertise to respond to changing circumstances can be acquired through experience and the evaluation of earlier episodes – not as anarchic as the kaleidic vision of Shackle.

Naturalistic representations also give much more attention to control, rather than to hypothetical gambles often associated with expected utility and other axiomatic decision theories which rely on probabilities. In looking at complex situations, especially dynamic and adaptive ones, such representations support views that are hostile to the normative promotion of classical theories – the normative standard being dismissed as either irrelevant since these theories misspecify realistic goals or questionable in the light of their poor characterizations of environmental constraints and perceptional frames, with the 'procrustean' elements of expected-utility theory, for example, unrecognizable as actual constraints for the participants and possibilities in any 'natural' settings.

Imprecision due to instability and change leaves little room for techniques which rely on stationary probabilities or stochastic processes. It may also severely restrict the value of laboratory experiments of the kind that question fundamental axioms of the EU theory (e.g. transitivity of preferences) because feasible experimental conditions might offer an inadequate approximation for the complex connections of real-world events and thus inhibit the drawing of valid inferences by economic agents. Naturalistic decision-making, therefore, may provide a challenge to experimental findings on cognitive biases and individual irrationality (which are considered in the next chapter), while it strengthens support for those attitudes which stress the diversity of form, context dependence and group-based choices – all relevant features which are awkward for integration with probabilistic notions and which are accompanied by real difficulties for testing or validating alternative models.

As with recent assessments of fuzzy methods, any true test of modelling and experimentation (whether based on case studies or on statistical analyses) requires the meta-evaluation of substantial applications rather than myopic academic debate. This evaluation has still to be produced, although references cited by Klein *et al.* (1993) give an initial glimpse of some actual and potential uses of naturalistic approaches, and they indicate a few basic themes in the controversial dialogue concerned with the acceptability of alternative research methodologies, including recent psychological experimentation and potential limitations on the generalizability of individual findings.

UNACCEPTABLE/ACCEPTABLE RISK

Naturalistic decision-making and the social judgement theories (Hammond *et al.*, 1975) point to causal ambiguity which affects the untangling of environmental relationships and weakens abilities to quantify risks of alternative complex activities. Such approaches focus on descriptive case studies for major economic and political decisions or judgements, which are made within small groups or with the assistance of 'expert' groups in risk evaluation. Generally, these 'high-stake' studies include unresolved conflicts, intuition, extrapolation and discretion as further sources of ambiguity. When moving beyond description, they often cite numerical indicators to guide effective judgements of vague acceptable (or unacceptable) risks and their estimation. A common feature of responses to the studies is persistent disagreement surrounding the validity of expert testimony and, especially, controversial dialogues on the relative accuracy of quantitative guidelines to any inherent risk. Clearly, despite the long experience with statistical modelling and project (or hazard) evaluations, there remains considerable scope for the promotion of alternative methodologies to provide tentative estimates of risk.

Despite the fashionability of expected utility and axiomatic theories in professional literature, it seems that such normative theories of rational choice are seldom invoked in applications of risk-appraisal techniques to major hazards. Similarly, the normal role of statistical inference based on imprecise probabilities is frequently qualified by an awareness of environmental ambiguity, biases stemming from acute fears of either catastrophic outcomes or 'unknown risk' (Baron, 1993: 172), qualitative diversity, multiple causation, and ethical slanting. The frequent presence of such qualifying elements means that 'probabilistic risk' evaluations of major hazards are essentially concerned with partial ignorance that cannot be sensibly expressed in terms of meaningful probabilities, subjective or otherwise – we have situations of ignorance rather than randomness and familiar probabilistic structures.

Lane gives one impression of the disturbed format for use of *emaciated* probabilities in some causality assessments:

> What the experts do is to pass in review the data they have, a process of 'anchoring and adjustment', to produce their evaluation of the required probability; that is, probability evaluations are arrived at by starting from available frequencies, ... adjustments are made based on differences [that] experts perceive between the observed and reference populations and attributes and on their judgements about the significance of these differences .... [The process] is dynamic, involving a continuous dialogue between the panellists' opinions, their data and the requirements of coherence.
>
> (1989: 98)

In practice, the situation is often much worse since data are sparse, there are no initial frequencies, panels of experts fail to reach a consensus because of incomplete information (and also causal complexities and imprecise goals), 'imagined' probabilities or frequencies are viewed sceptically, the 'unabashed activism' (Watson and Buede, 1987) of experts might induce hostile responses, populations shift endogenously, basic technical adjustments are not justified (or not justifiable), data are only available at aggregate levels, and the inferential outcomes or interpretations are affected by haphazard interactions with external issues, regulation and socio-political pressures (Kaser, 1989). This is far from the pristine simplicity which is usually sketched for most probabilistic models, whether aimed at retrodictive assessments or their predictive counterparts.

Although problematical and imprecise, major risk evaluations cannot be abandoned. They deal with important concerns for which systemic uncertainty might be reduced, but not by pretending that the necessary conditions of probability are fulfilled and that the interpretation of familiar statistics is unaffected. Instead, the problems imply a strong need to accept less precise answers and to change the rhetoric of probable inference to offer simply plausible explanations and support pragmatic conclusions. Fischhoff *et al.* (1982: xiii) offer a short list of recommendations – e.g. explicitly recognize inherent complexities, acknowledge the limitations of current approaches and expertise, develop better guidelines for conducting evaluations and reporting practices, encourage wider participation in the decision-making process and strengthen the public's understanding of hazards. We could add the need for more restraint on the active promotion of probabilistic models, which might be sound for simple well-defined problems but which are misleading and costly when applied without sufficient caution to complex ill-defined situations and then assumed to have normative or substantive significance without any basis other than a familiarity of technique.

Ambiguous environments lead to non-closed solutions, adaptive strategies, and a groping for conditional or tentative 'truths' – even if these are unpalatable and essentially vague, perhaps never to reveal underlying 'structures' (if there are any) with enough clarity. In this context, meta-analysis is not the combination of evidence within some comprehensive statistical framework, readily amenable to formal testing with probabilistic elements and subject to standard interpretation. Instead, future meta-analyses should give special attention to 'errors of the third kind', namely those errors arising from persistent failures to ask the right questions! Here, neither economists nor statisticians have very much claim to expertise because of subject-matter ignorance, causal ambiguity, and the inertial pull of immanent standards or hubris. However, given the importance of risk evaluations in many hazardous areas cited at the beginning of this chapter and in other applications (as reported in Arkes and Hammond (1986), Watson and Buede (1987), von Winterfeldt and Edwards (1986), for example), some re-casting of both formal and informal research methods is now called for. Probability is not inevitable. A flexible use of probabilistic notions may often be fruitful as part of an informative strategy if these are expressed in exploratory rather than confirmatory terms, and if past predilections and false prophets are resisted. The focus is then on ambiguity, ignorance, imprecision and vagueness, but not impotence or irrelevance.

5

EXPERIMENTS, DYNAMICS AND COGNITIVE ILLUSIONS

Most economic theories seem to come from the introspection and imagination of economists rather than from their observation of the world around them. When axioms or consequences of choice theories are confronted with external empirical information, there appear to be some significant disparities between these aspects of familiar theories and the perceived features of actual choices. Major disparities can be explained in terms of abstraction, integration of supplemental assumptions, systematic flaws in much empirical data, ill-chosen or unrepresentative contexts, insufficient rewards being offered to economic agents, a limited need for mere generic predictions and average tendencies rather than for any specific predictions, or additional factors and considerations. These explanations reflect alternative perspectives on methodology, the awkward interaction of theory and evidence, tainting of evidence and abstraction of theories, and common perceptions of the relative supremacy of theory versus evidence in situations of potential conflict or disparity.

Most expressions of methodological views are both muddled and confrontational. Frequently, assertions insist that only a new theory can effectively displace a flawed theory, and that abstract theories must be tested (beyond the normal requirement of logical consistency) by reference to the appropriateness and contextual richness of predictions based on them. Introductory textbooks can be overwhelmed by logical positivism and crude empiricism, while advanced textbooks for both economics and econometrics reveal sharp divisions between theory and its 'applications' – so progressive elements in the refutation and falsification of economic theories are lost in a preoccupation with theoretical detail and the pursuit of fashionable interests. Thus, it is hardly surprising that mature economists, coming from such a background, will continue to muddle along and have difficulty in dealing with economic puzzles (such as the Allais and Ellsberg paradoxes) and 'irrationalities' associated with the apparent violations of theoretical axioms.

One difficulty stems from the subjective and individualistic tone of microeconomic theories of choice and decision. We seldom have any systematic evidence on individual behaviour, especially on subjective aspects of mental states behind this behaviour, and tend to rely on aggregate data, vague notions of representative (perhaps average) individuals, single-minded and rational motivations, and misapplications of the so-called 'laws of large numbers' (Lipsey *et al.* 1994: 27). Awareness of these shortcomings and the growth of computational and experimental facilities have stimulated interest in finding some alternative means for confirmation of theoretical propositions through social experiments with a few participants in controlled environments or through large-scale econometric models focusing on various financial incentives (e.g. in relation to minimum wages, housing and access to medical care).

Clearly such experimentation is not limited to confirmation. Its outcomes may suggest some modifications and innovations which transform the form of current theories. Alternatively, potential benefits may be submerged by active hostility to their structural specifications and by the strong criticism of technical validity and generalizability – resistance, perhaps, being focused on the weak parallels linking any experimental subjects with their real counterparts.

In the following sections, we take up experiments in economics and psychology to explore both confirmatory and critical reactions to expected utility theories, often described in economic research as 'irrationality' when EU-axioms are questioned. Then we consider some of the newer approaches to dynamic choice which have emerged from experimental findings – e.g. regret, prospect and disappointment theories – as well as important contextual elements identified by psychological experimentation, including the various framing and cognitive biases, decision 'traps' and the ambiguities which may affect actual choices. We also identify generalizations of expected utility theories and note a recent attempt to introduce consequentialism to such theories. Since expected utility remains a fundamental ingredient of subjective probabilities (as promoted by Savage and others), the principal consequences of experiments will clearly affect the interpretation, coherence and acceptance of such probabilities for economic choice.

Looking back over forty years of experimentation and debate focused on the subjective EU-theories, Edwards (1992) suggests a new consensus has emerged among prominent decision theorists and experimentalists – one in which maximization of subjective expected utility persists as a normative principle but has been rejected as 'a descriptive model of the behavior of *unaided* decision makers' (italics added), with the recognition that it must now be replaced even though no existing candidate for description 'has a long lead over its competitors'. Perhaps the long search for a behavioural model of individual choice that is both normative and sufficiently descriptive should be abandoned. Then, experimentalists can turn their attention to more constructive efforts rather than continue to demonstrate the admitted impracticality of EU-theory and waste their limited resources. Unfortunately, the partial dismissal of EU-theory also challenges the preoccupation of some neoclassical economists with ubiquitous rationality, e.g. Lucas (1981), and leads to an evident need to clarify how much rationality is appropriate for dealing with complex situations (Thaler, 1987).

Some economic experimentalists, e.g. Roth (1986, 1987), distinguish various 'polar' dialogues which can characterize the integration of their efforts with those of others and clarify the motivation for experiments – speaking to theorists, searching for facts and whispering in the ears of princes! One dialogue is with theorists and considers theoretical predictions and any unpredicted regularities that emerge for unambiguous controlled experiments. The second dialogue is among experimentalists and focuses on the potential impact of variables which have not been directly treated by theorists (i.e. on basic exploratory experiments rather than confirmatory ones). The final dialogue involves communication with policy makers and thus requires giving much more attention to the characteristics of natural environments and to actual limitations on the options that can be adopted. In this context, there is a sharp disagreement between the views of experimentalists and those of the naturalistic school, briefly identified in our terse account of 'possibilities' for Chapter 4.

Basically, it seems clear that any success with experimental approaches requires the imposition of substantial control over the postulated structures, whereas the naturalistic approaches stress the inevitable imprecision of evolving complex situations and focus more attention on the need for description rather than any simple search for apparent regularities. Proponents of both types of approaches acknowledge that they have wider dimensions than the other conventional expressions of economic theories. Smith (1982) argues, for example, that the 'laboratory experiments are real live economic systems, which are certainly richer, behaviorally, than the systems parametrized in our theories' – representative of the tone normally adopted by advocates of experiments, and leading to a common awareness of the fundamental need for 'integrating theory, experimental design, and observation' beyond the familiar myopia of conventional theoretical approaches. Furthermore, 'experimentation changes the way you think about economics' (Smith, 1989: 152) by requiring the evolution of relevant economic analysis as expressed through 'concepts and propositions capable of being or failing to be demonstrated'.

The recognition of wider dimensions is, of course, both a source of strength and weakness. They reduce the simplicity of implied results by identifying relevant qualifications and special cases, and they also require the specification of supplemental assumptions which can attract hostile criticism. On the other hand, the issue of finding sensible domains of applicability is better focused and more visible. Experiments permit calibration of their parameters and they effectively encourage an interest in both robustness and meta-analysis, which are more difficult to integrate in econometric research, through the enhanced feasibility of replication.

EXPERIMENTS

Experiments and the design of analogue systems are relatively new in economics although attempts to develop them can be traced throughout the last half century, culminating in the 1990s with the establishment of a stable network of laboratories for experimental economics (many of which are listed by Friedman and Sunder (1994, app. V)), the promotion of reporting standards (*Econometrica*, July 1991), and the emergence of professional textbooks and substantial collections of readings – e.g. Davis and Holt (1993), Friedman and Sundar (1994), Hey (1991), Kagel and Roth (1993), and Smith (1991) – to consolidate the earlier surveys and major commentaries of, for example, Binmore (1987), Plott (1982, 1986), and Roth (1986, 1987). Experiments have been directed at methodological issues, market organization, conspiracy and collusion, financial assets, search, product quality and other informational questions, public goods, externalities, committee decisions, individual choice, games and a host of important topics in economics, political science, sociology and psychology. Some idea of the comprehensive scope of efforts in relation to a few of these items can be drawn from the list of readings provided by Plott in Friedman and Sundar (1994, app. I), or from entries in the subject index of the *Journal of Economic Literature* since 1986.

Economists, to some extent, are latecomers to experimental approaches. Their current interest in such approaches was clearly stimulated by earlier psychological experiments, which challenged the axioms of expected utility theory, and by various attempts to explore gaming perspectives – analogue models, such as those of Phillips and Strotz dealing with macroeconomic models (including leakages and non-linearities) in the 1950s, failed to generate any sustained interest in the benefits of unconventional approaches. Indeed, until the mid-1980s, the most prominent economic textbook in North America (Samuelson and Nordhaus, 1985) still dismissed the feasibility of any controlled experimentation to clarify economic relationships because of their inherent complexity and difficulties in controlling important influences.

Brief and incomplete outlines of the historical evolution of experimental economics are provided by Davis and Holt (1993, sect. 1.2) and Friedman and Sunder (1994, ch. 9). A relatively unknown effort of Chamberlin (1948) is now usually treated as one of the earliest reports of experimentation and a good illustration of failure to generate further research on experiments by economists. European origins are often linked to a series of papers written or edited by Sauermann, including a collaboration with Selten (1959) in the 1950s.

The primary concern for the interpretation of experiments and potential generalization of their findings stems from abstraction and the effectiveness of a search for appropriate parallelisms with the economic phenomena which are the focus of attention. Clearly, all microeconomic theories must abstract from many aspects of human activities and focus on a few factors that are viewed as especially interesting (significant or relevant) in some context. Experiments may provide a valuable means of substantially lowering the degree of abstraction, possibly to the extent that 'there is no question that the laboratory provides ample possibilities for falsifying any theory we might wish to test' (Smith, 1982 [1991]: 267).

The notion of parallelism (or transferability) for experiments generates attempts to justify a strong assertion that some useful propositions about the behaviour of individual economic agents and the primary impacts of institutional influences – as demonstrated or tested in recent laboratory experiments – can also be applied to many non-laboratory conditions when similar qualifications are satisfied, at least approximately. There are various sub-categories of parallelism, primarily depending on whether theoretic structures or their real counterparts offer a referential framework.

Initially, experiments might be designed to be close to some natural environments or, alternatively, close to the hypothetical conditions of economic theories (Plott, 1987), with design choices being affected by the main objectives of the experiments and by the dialogue anticipated for their findings. Ultimately, it is expected that comparison studies will seek to confirm experimental behaviour and verify the extent of parallelism that really exists, and thus stimulate another round of laboratory experiments and verification of theoretical components. Then, the availability of an evolving pattern of substantial experimental evidence might provide an effective framework for the ongoing interpretation of research findings, while establishing the relative value of continuing to pursue the experimental method itself.

An optimistic and promotional account of this dynamic process for experiments is provided by Smith (1989), who insists that much experimental evidence is consistent with the familiar predictions of our market theories in repetitive settings, despite a persistent awkwardness in reconciling the normal assumptions of individual rationality with the cognitive illusions, myopia and computational limitations of real economic agents. To assist this reconciliation and to diminish the impact of apparent irrationality, Smith (1985) suggests that some violations of EU axioms might simply reflect the need for a better treatment of transactional costs in the formation of economic theory, involving the subjective elements of 'thinking, calculating, deciding, and acting' for example, which can also blur the distinction between normative and positive theory.

Another tentative form of reconciliation, less supportive of experimental results, is offered by Binmore who insists that routine replication is a basic requirement for experimental economics:

> Until a piece of work has been replicated elsewhere by others, preferably several times, it cannot be regarded as properly established.
>
> (1987: 260)

He also points to the possible failures of both psychologists and economic experimentalists to recognize the 'unspoken caveats' which many theoretical economists take for granted, besides repeating an earlier view that the individual reliance on heuristics may reflect an outcome of some adaptive process, even an optimizing process, since:

the fact that non-optimizing behavior may be engineered in the laboratory does not necessarily mean that the behavior is not optimal in the environment to which it is adapted.

(Binmore, 1987: 263)

However, this latter suggestion seems to stretch our credulity in retaining the applicability of current theories. A contrary view on the complementarity of rationality and adaptation is provided by Lucas (1986), a prominent economic theorist who actively stressed equilibrium modelling and ubiquitous rationality.

A fundamental difference between the data used in econometrics and those of most experiments arises in connection with the ease of replication. Most of the common econometric data can be described as 'happenstance', developed for other environmental purposes which are distinct from the contexts assumed by many economic researchers (e.g. standard indices for national income accounts or consumer price indices). These data are usually subject to timing, conceptual, aggregation, measurement and other significant flaws. On the other hand, laboratory data are controlled and synthetic, being created to simulate hypothetical conditions and presumed to be readily replicable in a way that few happenstance data can be replicated (Binmore, 1987: 258). In particular, the real data are either non-replicable or too costly to replicate for econometric analyses. The major flaws of experimental data, on the other hand, reflect the limited knowledge or imagination of investigators in dealing with structural specifications of the hypothetical context, perhaps dominated too often by the usual simplicity of fashionable economic theory rather than by a progressive reduction of the main conditioning influences that are active in our complex situations.

It should also be realized that replication, by itself, does not establish the relative accuracy of experimental findings as a reflection of real phenomena. As Sutton (1987: 284) recognized in his review of a wide range of findings for bargaining experiments, their inherent limitations, restricted scope and fragile evidence may be restrained by further experimentation, and thus serve to 'counterbalance the worry that the "real world" which gives us good data so grudgingly, may always be too poorly mimicked by the interactions of our experimental subjects'. In any case, successful parallelism has yet to be clearly demonstrated, since experiments and the von Neumann–

Morgenstern models still lack descriptive power in a number of important respects.

EXPERIMENTAL IRRATIONALITY

The recent emergence of experimental options and the promotion of interesting behavioural theories by economists and psychologists to challenge earlier rationalist preoccupations (e.g. as identified with the EU-axioms and rational expectations) will not lead to a clear theoretical contest from which one approach emerges to dominate its rivals. Contrary to some visions of methodological progress, the rival approaches are incommensurable and their future discussion might better address various perceptions of 'comparative advantages'. Thus Zeckhauser, perhaps with weary recognition, points out:

> for any tenet of rational choice, the behavioralists ... can produce a laboratory counterexample, [while] for any 'violation' of rational behavior discovered in a real world market ..., the rationalists will be sufficiently creative to reconstruct a rational explanation, [and] elegant abstract formulations will be developed by both sides, frequently addressing the same points, but because there are sufficient degrees of freedom when creating a model, they will come to quite different conclusions.
>
> (1986: S438)

This is an uncomfortable but valid impression of economic research. We are left with the incomplete task of describing the experimental findings without being sure about their implications, and equally unsure about the relevance of models assuming strong rationality by decision-makers. By extension, subjective expected utility and the probabilistic concepts based on it suffer from ambiguity – can such concepts still be derived from frameworks which 'generalize' the axioms initially adopted by Savage and other subjectivists?

Interest in experimentation was strongly stimulated by the reports of non-economists which cast doubt over the basic EU-axioms that are described in Chapter 2, including the assumptions of preferential transitivity and independence. Experiments discussed in the *Journal of Experimental Psychology*, such as those of Lindman (1971), Lichtenstein and Slovic (1971), and elsewhere (Tversky, 1969) were later revived by Grether and Plott (1979) and others in the economic literature. Currently, there are several influential collections of important papers, including Kahneman *et al.* (1982), Arkes and Hammond (1986) and

Gardenfors and Sahlin (1988), and a host of individual papers – e.g. those of Plott (1986), Slovic and Lichtenstein (1983), Tversky *et al.* (1990), Karni and Safra (1987a, b), Appleby and Starmer (1987), Schkade and Johnson (1989) and Bostic *et al.* (1990) – which created a rapidly expanding area of research interests that offers a rich menu for characterizing individual choice and decisions, generally beyond the restrictive framework of EU-models. In 1982, the survey by Schoemaker indicated just nine variants of EU-models. This number increased markedly during the subsequent fifteen years.

It is tempting to treat relevant experiments as a whole rather than separating them according to whether they stem from research by economists or from the efforts of psychologists. Experiments are relevant if they help us understand some aspects of behaviour and their corresponding mental states and processes – e.g. choosing, deciding, preference, recalling and reflecting. All experiments (laboratory and real-life) must involve abstraction and artificiality so that inferences derived from their outcomes are always tentative, but this is a qualification for the confidence attached to *all* evidence in scientific research and we have a well-established literature on the basic issues of validation, including Campbell and Stanley (1963), Cook and Campbell (1976, 1979) and Overman (1988, Parts 2 and 3).

With psychological experiments, the degree of artificiality may be disturbing and the conditions for individual experiments may be found unsatisfactory. De Finetti, for example, notes:

> When the experiment is artificial, the problem for the subjects is ... to divine what the experimenter intended to test. This situation is like that in an amusing science-fiction novel, where an extraterrestrial animal reacted strangely in a psychological experiment because ... it was interested in and able to detect the psychology of the experimenter.
>
> (1974: 20)

Given this observation and similar ones, the discovery of 'biases' in behaviour and preferences by psychologists is affected by our awareness of potential flaws in the experimental framework itself, but this is hardly unique to psychology. De Finetti is misleading when he insists there

is a strong distinction between what is objective, namely the observed facts, and what is not, namely, the beliefs about their interpretation and significance.

(De Finetti, 1974: 21)

Are facts ever objective when we are dealing with complex human situations?

Other interpretative difficulties are much less severe. For example, Holt (1986), Karni and Safra (1987a, b) and Segal (1988) suggest that preference reversals might stem not from violations of the transitivity axiom, but rather from a violation of other axioms (perhaps that of independence), and there are many complaints that the designs of experiments contain inadequate financial rewards for their subjects and thus weaken their motivation, affecting observed outcomes. Camerer (1987) offers a brief list of the major defences of conventional approaches in the face of strong criticisms derived from experimental challenges to EU-axioms.

HEURISTICS AND BIASES FOR INTUITIVE PROBABILITY

The basic conflict between major experimental results and the rational Savage-like bases for subjective probabilities is clearly identified in two generalizations expressed by Kahneman and Tversky, and is not simply a matter of some potential violations of the EU-axioms:

> people do not follow the principles of probability theory in judging the likelihood of uncertain events . . . hardly surprising because many of the laws of chance are neither intuitively apparent, nor easy to apply.
>   . . . the deviations of subjective from objective probability seem reliable, systematic, and difficult to eliminate. Apparently, people replace the laws of chance by heuristics, which sometimes yield reasonable estimates and quite often do not.

(1972: 25)

The alternative focus on heuristics (or rules of thumb) and the new axiomatizations of generalized expected utility, both influenced by the experimental findings and some creative reactions to paradoxes, provide a very different environment for exploring the economics of choice and decisions – one that is not just made up of some small concerns (putting epicycles on a Ptolemaic vision of the Universe), but rather a shattering of the 'old time religion' and an effective demise of

100

our former representations of the rational economic man (Edwards, 1992a).

The heuristic focus might also provide a better descriptive framework for the elucidation of subjective probabilities, since it gives more attention to the dynamic and mental context out of which such probabilities emerge, are retrieved, or are revised. This descriptive framework, as being elaborated in recent experimental findings, appears to discourage the view of 'man as a reasonable intuitive statistician' and casts doubt on individuals' skills to assess the effects of sample size, conditioning factors and other contextual features. Some illustrative images can be developed by the consideration of two particular heuristics (for mental effort underlying probability) noted by Tversky and Kahneman (1971, 1973), 'representativeness' and 'availability', and by their account of a perverse law of 'small' numbers. The development of these elements stresses relative success in determining probabilities rather than retaining the normative standard of our historical literature.

Observations suggest that an individual's intuitive evaluation of a firm probability for some uncertain event might involve the degree to which this event is 'similar in essential properties to its parent population' and to which it reflects 'salient features of the process by which it is generated':

> in many situations, an event A is judged more probable than an event B whenever A appears more representative than B. In other words, the ordering of events by their subjective probabilities coincides with their ordering by representativeness.
>
> (Kahneman and Tversky, 1972: 26)

Initially, this heuristic appears somewhat vague as a common form of avoiding cognitive strain. It is clarified and given more precision by Bar-Hillel (1980a, 1982) and Tversky and Kahneman (1974, 1982).

The basic notion behind the availability heuristic is that a probability will often be partially judged by ease of recollection and imagination:

> the number of relevant instances that could be readily retrieved or the ease with which they come to mind are major clues that men use in estimating probability or frequency.
>
> (ibid. 44).

Popular interest in this proposition was encouraged by Tversky and Kahneman (1973), who illustrated its potential occurrence with a series of case studies. Other aspects are discussed by Ross and Sicoly (1979), and by Taylor (1982) who suggests that due attention to availability reveals informational processing errors that can be understood without recourse to motivational constructs, permits the identification of cognitive biases which explain blatant deviations from rational theories, and facilitates the description of dynamic corrections that diminish the incidence or magnitude of deviations.

Discussions of a perverse law of small numbers focus attention on some important inferential issues, reflected in the behavioural assertions of Tversky and Kahneman:

> that people have strong intuitions about random sampling; that these intuitions are wrong in fundamental respects; that these intuitions are shared by naive subjects and by trained scientists; and that they are applied with unfortunate consequences in the course of scientific inquiry.

> (1971: 106)

These assertions, if correct, establish a basis for believing that there are frequent biases for intuitive judgements of probability, especially in regard to the perceived stochastic properties of any statistics drawn from random samples – biases which psychologists have attempted to enumerate, perhaps too successfully (Kahneman, 1991)! Tversky and Kahneman suggest that a believer in the law of small numbers will (in good faith) overestimate power and significance, underestimate the width of confidence intervals, and not correct his mistakes:

- he gambles his research hypotheses on small samples without realizing that the odds against him are unreasonably high;
- he has undue confidence in early trends and in the stability of observed patterns;
- in evaluating replications, his and 'others', he has unreasonably high expectations about the replicability of significant results;
- he rarely attributes a deviation of results from expectations to sampling variability, because he finds a causal 'explanation' for any discrepancy in action. Thus, he has little opportunity to recognize sampling variation in action.

> (ibid. 108)

These are substantial flaws which cannot be eliminated by ignoring the misconceptions that they describe. A valuable account of the intrusive new views on human cognitive processes and their direct relevance for the assessment of subjective probability is provided by Hogarth (1975), who suggests that man as 'a selective, step-wise information processing system with limited capacity, is ill-suited to the task of assessing probability distributions within the framework of the more common statistical models'.

A clearer recognition of heuristics and their enumeration does not exhaust the potential impacts of recent psychological research on subjective probability theory or on its association with various notions of expected utility and coherence. It also seems clear that actual evaluations or assessments of numerical probabilities will be affected by task characteristics, meaningfulness, feedback mechanisms and other environmental influences. More attention can be given to such elements, including a more systematic approach to experimentation which explores and clarifies their roles, without reducing the strong appreciation of man's successful performance in a large number of complex areas involving uncertainty and other forms of imprecision.

Heuristics and the associated biases now provide an important correctional force for past treatments of probability in relation to choice and decisions. Focusing on actual performance, learning from experience, and processes for coming to a decision shifts the character of subjective probability away from the orderliness and pristine simplicity of static rationality to a more action-oriented and dynamic perspective view, acknowledging both context-dependency and severe cognitive limitations, but excluding an informal reliance on vague and comfortable notions such as efficiency, equilibrium and evolution to support arbitrary normative standards. However, most economists are not precluded from pursuing any theory with unrealistic behavioural assumptions, for they can always, as noted by Lopes (1994: 216) 'accord the theory "as if" status and get on with the business of using the theory as a heuristic device for exploring reality' and thus display (or confront) the mysterious power of markets! On the other hand, it might seem appropriate to offer more assistance to actual decision-makers so that they can avoid costly misconceptions or biases and thus reduce the potential obstacles to efficiency, as demonstrated by Russo and Schoemaker (1989).

On a wider scale, the recognition of heuristics may indicate a basic fissure in cognitive science:

On the one side we have an approach that starts with language and logic and that views thinking as a process of inference and reasoning, usually using a language-like representation. On the other hand we have an approach that views thinking (especially problem solving and concept attainment) as a process of heuristic search for problem solutions, generally using representations that model, in some sense, the problem situation.

(Simon and Kaplan, 1989: 14)

For economists, the primary issues are whether logical reasoning (presumably characterized by the familiar axioms and coherence) or a multi-stage search (involving both heuristics and mental models) is 'the correct metaphor for thinking', and whether both approaches to thinking have merit or are interconnected. In the future, such issues will be enlarged as the current preoccupation with cognitive factors in judgemental biases and choice is supplanted by wider perspectives – irrationality is not simply a failure of reasoning (Kahneman, 1991; Lopes, 1994), since social heuristics, passion and other elements might be influential too.

CONTEXT: AMBIGUITY, FRAMING AND OTHER BIASES

Many elementary courses in statistics illustrate the potential outcomes of tossing a fair coin and of drawing a coloured ball from an urn with a known proportion of contents. However, these courses seldom deal with alternative situations in which the coin is biased or the urn's contents are unclear. Such fundamental ambiguity, due to incomplete probabilistic information, was initially stressed by Ellsberg (1961) when he questioned the routine acceptance of EU-axioms for rationality. Savage accepted the potential for numerical probabilities to be 'unsure' and he considered the view that higher-order probabilities (i.e. probabilities about probabilities) might exist, but then dismissed the practical value of this extension:

> There is some temptation to introduce probabilities of second order .... But such a program seems to meet insurmountable difficulties .... [Once] second order probabilities are introduced, the introduction of an endless hierarchy seems inescapable. Such a hierarchy seems very difficult to interpret, and it seems at best to make the theory less realistic, not more.
>
> (Savage, 1972: 58)

More recently, Levi (1986: 29) argued that most paradoxes and any apparent deviations from the maximization of expected utility 'can by and large be understood' as located in situations with no legitimate numerical probabilities or lacking utility assignments that are unique up to an affine transformation – suggesting that they are effective criticisms of Bayesian dogma, persistent in the face of ignorance and indeterminacy, rather than simply attacks on subjective probability itself. An alternative approach is offered by Schmeider (1989), who retained the maximization of expected utility, but based its evaluation on some non-additive probabilities which can represent the transmission and recording of information and amendments of the normal EU-axioms.

A further discussion of Ellsberg's ambiguous probabilities or 'vague uncertainties' is offered by Wallsten (1990). He stresses the overwhelming need for enough information to support numerical magnitudes, especially in forecasting, while recognizing the value of linguistic representations of probabilities in vague situations. With regard to imprecise utility assignments or preferences, Butler and Loomes (1988) point to the potential occurrence of a 'sphere of haziness' which can be attributed to the inherent imprecision of both information and preferences, incommensurable choices, and the corresponding awkwardness in calculating certainty equivalents for risky prospects.

More ambitious responses to the recognition of ambiguity and the hazards of unreliable probabilities are provided by Einhorn and Hogarth (1986), Fishburn (1991, 1993), Frisch and Baron (1988) and Gardenfors and Sahlin (1983). The reliance on expected utility for modelling choice and decisions in the presence of uncertainty is disputed by Einhorn and Hogarth on three substantial grounds. They question the common metaphor of choice among explicit gambles and lotteries as unrealistic since gambling devices (e.g. dice and urns) do not adequately represent the principal characteristics of actual uncertainty facing economic agents. They also insist that decision makers are 'highly sensitive to contextual variables' and that 'changes in context strongly affect the evaluation of risk'. Finally, they suggest that potential payoffs can 'systematically affect the weight given to uncertainty, especially in the presence of ambiguity', so the presumed independence of probabilities and rewards is dubious. Given this background, Einhorn and Hogarth clarify the nature of ambiguity, its avoidance as identified by Ellsberg, and its attractiveness using a model in which economic agents anchor their initial probabilities (with prior information or experts' guesses) and subsequently modify them to reflect the untidy world of ambiguity. What emerges from their discussion is a firm assertion that it would be foolish to ignore the ambiguous nature of probabilities, contextual and framing effects, regret, the illusions of control, superstitions, and the dependencies of probabilities and utilities by remaining tied to familiar axioms, which are based on gambles and lotteries.

Fishburn takes a different innovative approach in seeking to provide a set of new axioms to represent the comparative ambiguity relation, without conclusively demonstrating that the primitive concept of 'event ambiguity' adds much to our existing treatment of comparative probability (Fine, 1973), or that it can be explored empirically without reference to likelihood or choice. Gardenfors and Sahlin, by contrast, opt for a focus on the interesting notion that 'the information available concerning the possible states and outcomes of a decision situation has different degrees of *epistemic reliability*'. This focus permits them to explore various deviations from standard Bayesian theories (e.g. as associated with de Finetti and Savage), to deal with potential unreliability and the non-uniqueness of probability distributions due to an inadequate quality of knowledge.

Drawing on earlier literature, Frisch and Baron (1988: 152) note that individual probability judgements may not sum to unity when a situation is ambiguous: most experimental subjects prefer non-ambiguous gambles except at very low probabilities (where there might be a preference for ambiguity) and ambiguity causes people to be unwilling to act. However, they do not see this unfortunate background as opposed to more reliance on subjective probabilities. Instead, they suggest that ambiguity simply directs attention to the important limitation whereby optimal decisions are prescribed only relative to what is known! This view seems disingenuous.

The rich variety of contexts for making choices and decisions, elucidating useful information, determining firm preferences and estimating probabilities contain many important features beyond those which we have already identified with ambiguity and the two heuristics of representativeness and availability. It is convenient to recognize a 'decision frame' (Tversky and Kahneman, 1981), which includes 'the decision-maker's conception of the acts, outcomes, and contingencies associated with a particular choice', and also to acknowledge that this frame is partially controlled by the way in which any problems are formulated as well as by 'norms, habits and characteristics of the decision-maker'. In addition to the framing effect (Tversky and Kahneman, 1986), substantial attention needs to be given to other heuristics and corresponding biases as expressed in processes or mechanisms, generalizations and effects put forward over the last two decades – all of which may seriously affect the integration of probability for the economic models of choice and decisions.

These important supplements include loss aversion or asymmetry in the valuation of losses and gains, status quo bias (Samuelson and Zeckhauser, 1988), flawed hindsight or the tendency to view what has already happened as now relatively inevitable and obvious (Fischhoff and Beyth, 1975; Fischhoff, 1980), excessive optimism and over-confidence (Oskamp, 1965), conservatism (Edwards, 1968), sunk costs (Thaler, 1987), and base rates (Ajzen, 1977; Bar-Hillel, 1980a, b, 1990; Fischhoff and Bar-Hillel, 1984). Easy access to the relevant literature on such supplements is assisted by collections of major papers assembled in Arkes and Hammond (1986) and Kahneman *et al.* (1982), and by brief summaries such as that of Evans.

DYNAMIC CONSIDERATIONS AND OTHER MODIFICATIONS

Much of the recent elaboration of heuristics and cognitive biases still comes from experimental psychologists rather than from analytical economists, but the main reassessments of the formal EU-frameworks now occur within a multi-disciplinary environment where the re-search findings are shared and quickly assimilated, and where sub-stantial cross-fertilization is common. Increasingly, a dynamic slant is being added and the initial axioms for rational choice or actions are being generalized. Beyond the old paradoxes and radical treatments of both imprecision and possibilities discussed in our earlier chapters, three strands of development during the 1980s seem especially important for the future acceptance of modified EU-frameworks as normative and descriptive standards. The first strand involves novel descriptive theories associated with the editing of prospects (Kahne-man and Tversky, 1979), a new perspective index influenced by Shackle's imaginative approach (Ford, 1987), the contaminating effect of anticipated regret (Bell, 1982; Loomes and Sugden, 1982; Fish-burn, 1982), and disappointment (Bell, 1985; Loomes and Sugden, 1986). The second strand focuses on potential generalizations of EU-theory, primarily in response to earlier criticisms, as surveyed by Machina (1983, 1987, 1989), while the final strand introduces a new form of consequentialism to dynamic variants of the EU-theory (Hammond, 1988a, b; McClennan, 1990). Obviously, this brief list suffers from omissions – other relevant developments can be traced through the efforts of Edwards, Luce and Herrnstein, as well as the prominent authors already cited above.

Prospect theory adds a multi-stage character to the treatment of choice and decisions, separating a preparatory phase of framing and editing from a subsequent phase of evaluation. Preparation includes a framing of acts, contingencies and outcomes as mediated through norms, habits and expectancies. It will be accompanied by some cancellations and elimination of dominated options, then followed by an evaluation of remaining prospects and the selection of a favoured choice from among them. Any clarification of this extended process can draw on earlier research findings and hence make 'decision-makers' seem more realistic, thus emphasizing the transparent descriptive tone of prospect theory by assuming limited information-processing ability, major cognitive biases (including some misconceptions of probabilities, perhaps involving non-linear transformations to connect subjective weights with the inherent objective probabilities), and frequent reliance of convenient rules of thumb. Kahneman and Tversky (1979) developed this approach to include perceptional (certainty, reflection and isolation) effects which, with other ingredients, influence the realistic slanting of decision weights and a value function (somewhat removed from the normal probabilities and utility functions of the EU-theory).

The emergence of theories involving regret and disappointment reflects a wider vision of utility which weakens the acceptance of Savage's sure-thing principle and the assumption of transitivity as representations for rationality. Utility is associated with the dynamic awareness which comes from remembering historical events and anticipating new ones. Current choices and decisions recognize the 'importance of what might have been' (Loomes and Sugden, 1984) and are not isolated in time – quite different from the atemporal context of von Neumann and Morgenstern, which may have diverted economists' attention from sequencing and other dynamic phenomena (Pope, 1986) as well as muddled interpretation of the Savage–Allais exchange in the early 1950s.

The new regret and disappointment theories also challenge the more complicated prospect theory (Loomes and Sugden, 1982; Loomes, 1988b), although clearly stimulated by the same violations of EU-axioms and searches for alternative models of optimality. Instead of a preoccupation with 'systematic violations', notions of regret give less attention to behavioural obstacles and simply amend the usual cardinal utility function by adding a new component

(Sugden, 1985: 172) to reflect the disappointment or elation associated by an individual economic agent with each action (thereby introducing a degree of non-separability across consequences). This yields an expression for expected modified utility and offers a much simpler means of resolving some potential paradoxes, but still represents a fundamental break with the conventional EU-framework. A dynamic aspect is embodied in the presumption that an individual will learn by experience and thus be able to predict experiences of regret and rejoicing. Modifying utility in this way explicitly undermines the view that EU-axioms are self-evident propositions about rational choice with general validity.

When von Neumann and Morgenstern began the transformation of utility theory, there was confusion over the inclusion and merits of particular axioms (especially the independence axiom) before later experimental evidence emerged to challenge the presumption of transitivity. Discontent with the independence axiom has continued to stimulate attempts to generalize expected utility by modifying or excluding this particular axiom and by introducing various forms of non-linearity for preferences and probabilities – as illustrated by Machina (1983) – while retaining some of the initial elegance of earlier versions of EU-theory. Recent non-expected utility models of preferences, their dynamic consistency, and tests of them are summarized in Machina (1987: 132–6; 1989), Weber and Camerer (1987), and Camerer (1992). An interesting reaction to Machina's own 'neo-Bernoullian' efforts is provided by Allais (1984).

The final strand of recent developments in this area, which introduced a new form of consequentialism to EU-theory, has failed to generate much attention from either economists or psychologists, perhaps because it seems to lack empirical interest. This lack of attention is unfortunate since it means that a fundamental problem is being relatively ignored. Many choices and decisions arise in multi-stage or sequential contexts, which are spread over time. In making a choice or decision, perceived benefits and costs should be evaluated in terms of the events and possibilities at other stages. Textbooks on decision-making – such as von Winterfeldt and Edwards (1986, ch. 3), Raiffa (1968) and Watson and Buede (1987, ch. 3) – typically express multi-stage situations with 'decision trees' for which given values of probabilities and payoffs are specified along each branch.

A reliance on such trees is clearly effective when all dynamic consequences are known and ineffective otherwise. The existence of extensive knowledge reduces dynamic choice to simple calculations and rationality can again be linked to EU-axioms. Recently, Hammond has sought to clarify principles of consequentialism which might be useful in the characterization of dynamic or sequential choice. He argued that many apparent violations of familiar axioms might stem from myopic failures to consider the full consequences of potential actions (i.e. from the use of mis-specified decision trees), detecting the corresponding occurrence of dynamic inconsistency. When these failures are eliminated, he suggests, consequentialism leads to intuitively-acceptable set of principles which resemble their static counterparts. McClennen (1990) and LaValle (1992) are less sanguine and dispute the plausibility of Hammond's arguments.

When decision trees are underspecified or imprecise, their implementation is as difficult as finding a normative path through heuristic searches or generalized EU-theories. Major problems stem from the recognition that actual sequential processes seldom retain a fixed horizon, that potential options change and become clearer (and 'rethinking' becomes more attractive) while the processes are under way, that iterative exercises are common, and full enumeration is never costless. Thus, we are driven by such complexity away from the analytic elegance of EU-frameworks and towards the naturalistic approaches identified at the end of the last chapter, and hence may opt for a better description of human actions instead of the academic pursuit of normative models and standards.

6

EVIDENCE, SUPPORT
AND BELIEFS

The multi-faceted character of probability has been visible since the time of Pascal (Hacking, 1975). At one extreme, there are statistical or frequentist conceptions that lead to the current foundations of standard econometrics, ergodicity, stationarity and testing for statistical significance. At the other extreme, there are treatments of reasonable degrees of beliefs, epistemic states, foundations of expected utility theory, and various subjectivist or personalist conceptions. In the present century, with the gradual recognition of degrees of belief, refinements of rationality, and new awareness of cognitive biases (with more attention being given to the basic issues of support, information and evidence), primary views of probability allow a considerable range of measurability and wider environmental contexts in the consideration of dynamic uncertainty.

The range of measurability – reflected in the acceptance of less precision, limitations stemming from changed circumstances, and general awkwardness for the elucidating, sharing and revising of probabilities – is the main focus of the following discussion. There are practical difficulties in determining any probabilities, whether as limiting values for some infinite sequences of relative frequencies or as the credible and firm values lurking in people's minds. We should also note the strong suspicion that some newer concepts (from Shackle's myopic notion of potential surprise in the 1940s to its present counterparts, including belief functions and epistemic reliability) are remote from earlier probabilistic bases and may perhaps be essentially non-probabilistic. Fine illustrates the recent lack of enthusiasm for characterizations of chance and uncertainty – a lack of enthusiasm which is encouraged by fashionable reliance on arbitrary assumptions to complete simple procedures of statistical inference and decision rules (e.g. invariance, admissibility, linearity, rational expectations, unbiasedness and asymptotic efficiency), and which is also stimulated by substantial distaste with controversy arising out of theoretical confusions (due to the rival conceptions of probability rather than disagreements over rival techniques).

> The many difficulties encountered in attempts to understand and apply present-day theories of probability suggest the need for a new perspective. Conceivably, probability is not possible. A careful sifting of our intuitive expectations and requirements for a theory of probability might reveal that they are unfulfillable or even logically inconsistent.
>
> (Fine, 1973: 248)

Fine, somewhat radically, asks why we should not now ignore 'the complicated and hard to justify' normal structures for probability and statistical inference and proceed directly to 'those, perhaps qualitative, assumptions that characterize our source of random phenomena, the means at our disposal, and our task?'.

Since processes for the acquisition of relevant information by economic agents are seldom discussed in common models of economists (despite vague references to transaction costs and opportunistic search guided by reservation wages), it seems worthwhile to begin with an adaptive and evolutionary perspective of price-signalling which was promoted by Hayek and Polanyi to represent what actually occurs in price-driven and heterogeneous market conditions. They stress the tacit, incomplete and qualitative character of personal information, acquired through direct participation in markets and through spontaneous reactions to an erratic stream of exogenous and persistent novelties. This view of diffuse information arising in a non-stationary environment insists that 'knowledge exists only as the knowledge of individuals' and 'the great problem is how we can all profit from this knowledge, which exists only as the separate, partial, and sometimes conflicting beliefs of all men' (Hayek, 1960: 24–5). It is the main subject of our first section below.

Numerical imprecision is considered in the second section by reference to some major attempts (e.g. by Keynes, Hicks and Allais) to reconcile incomplete and qualitative probabilities with the demands of economic analysis, and more generally by reference to the mathematical treatments by Koopman, Dempster and Good which stress upper and lower bounds of imprecise probabilities and seek to determine appropriate axioms to represent qualitative concepts. This is followed by a brief comment on the 'weight of evidence' and the distinction between information and evidence, as clarified by Keynes and Good but anticipated by Peirce many years earlier.

We then turn to recent attempts to deal with dynamic beliefs and their revision in the human environments subject to exogenous shocks and other forms of novelty. 'Changing one's mind' seems to be a reasonably common occurrence in such environments. Here the Dempster–Shafer theory of belief functions and the novel modelling approach of Gardenfors to the dynamics of epistemic states provide convenient foci, although we could have drawn on Shackle's kaleidic vision and on naturalistic accounts of actual decision processes, which offer less formal but still interesting accounts of dynamic beliefs, their transitory nature, and their revision over time and changing circumstances.

The final two sections explore the consolidation, elucidation and confirmation of personal probabilities as well as the potential integration of their individual values in collective probabilities to be held by 'groups' of planners or decision-makers after short interactive processes of Bayesian dialogue and consensus seeking. In this context, subjectivists have frequently been criticized for relying on excessive precision in their analyses – what might be called the 'Bayesian dogma of precision' (Walley, 1991: 3, 241) – and for not giving enough attention to the practical aspects of any 'hard' choices where conflict within groups is unresolved (Levi, 1986b), a degree of pluralism is encouraged, and some non-pressing judgements are suspended and ignorance prevails.

Pratt firmly disputes the suggestion of excessive precision here:

No sensible person really thought that probability and utility assessments preexist in anyone's mind, or that probabilities of all events could or should be assessed directly and then checked for consistency, or that they be naturally consistent.

(1986: 498)

This strong opinion is surely a misrepresentation of the history of probability, both among economists and elsewhere. However, the term 'sensible' can be interpreted so widely as to exclude much of the previous statistical literature and create a fabulous past! Pratt is avoiding the real problem that 'evidence for probabilities may be inadequate or altogether missing' (Shafer, 1986: 500), so Savage's axioms for subjective probabilities and the EU-theory need to be extended, modified or discarded. Imprecision is not merely a matter of the usual posterior distribution of Bayesian analysis being sensitive to

113

choices of priors; i.e. it is not merely a small difficulty to be overcome by finding more data to sharpen the choice of priors.

PERSONAL KNOWLEDGE

From his initial paper on knowledge in 1937 to the end of a long professional career, Hayek stressed the notion that economic phenomena should be explored as part of an ongoing dynamic process in which significant knowledge and information lead to an evolution of individual understanding. He considered the effectiveness of competition to be driven by discovery through participation, with the hazards of ignorance and uncertainty partially resolved while some intentions and expectations remain unrealized. Knowledge is diffuse in markets, particular facts are transitory, many purposes are specific and temporary, and the eventual results of competition are both 'unpredictable and on the whole different from those which anyone has or could have deliberately aimed at' (Hayek, 1978: 180). In this vision of economic conditions, prices serve as a monitoring mechanism for the individual economic agents, by directing their 'attention to what is worth finding out about market offers' and by encouraging 'a capacity to find out particular circumstances, which becomes effective only if possessors of this knowledge are informed by the market which kinds of things or services are wanted, and how urgently they are wanted' (Hayek, 1978: 182).

Within this kaleidic society, all economic adjustment is made necessary by unforeseen changes and much knowledge originates from events that are exogenous to individual agents. What matters here is an evolutionary emergence of qualitative changes or 'novelties'.

As the simile for actual processes, this vision severely restricts analytical models and quantitative prediction beyond that of broad patterns. It clearly leaves little room for numerical or symbolic probabilities found in conventional search theories or associated with expected utility, and the overall efficiency of markets (to the extent that it occurs) stems from the general responsiveness to price signals which facilitate an unforeseen coordination of the individual agents (Hayek, 1986).

In modern capitalist economies, this Hayek vision of adaptive behaviour and knowledge acquisition is somewhat flawed. It ignores common features of information such as the endogenous generation of information within larger firms, the privatization of internally-generated information, and the management of innovation (including research and development), which are seldom spontaneous. However, it still contains some relevant aspects of the complexity of human structures and their market environment. For more than fifty years, Hayek suggested that economic success came from human action but was not the result of human design, and that inherent contrariness may disturb the connections between individual plans or designs and the realized outcomes of economic behaviour. Thus, rationality has to be defined in a markedly different way, which acknowledges the substantial benefits of proximity (inhibiting centralized planning) and the limiting capacity of human minds for pattern recognition and intuition.

A reliance on familiar statistical methods (more generally, on the notion of sampling from a fundamental 'population') to reduce any misunderstandings of complex phenomena might also be irrelevant if qualitative changes are frequent, spontaneous and substantial, and also if 'it is the relations between individual elements with different attributes which matters' (Hayek, 1964: 30). Thus, any statistical predictions are limited to simple generic statements or patterns of little value in the detailed design of appropriate actions (Paque, 1990). The Hayek vision (especially, its rejection of shared or centralized knowledge and planning) gained prominence among economists with the arrival of conservative governments in the USA and Britain during the last two decades. However, its firm reliance on spontaneity, novelty and individualism makes it inconsistent with all of modern econometrics and much of the fashionable equilibrium-based micro-economic theory. Both macro-econometrics and the stochastic equilibrium theory are incompatible with Hayek's own abandonment of probability and aggregate data, and with his account of rational market processes and the acquisition of new information by individual economic agents. Anti-Keynesians cite him in their condemnation of governmental intervention and the corresponding 'distortions', while ignoring the awkwardness of the Hayek vision for all of the familiar stochastic processes.

INCOMPLETE AND QUALITATIVE PROBABILITIES

Other prominent economists have also moved away from density functions and measurable probability toward a deeper recognition of incomplete personal knowledge, rational degrees of belief, neglect or avoidance of small probabilities, and qualitative formulations of vagueness or imprecision. Such aspects introduce considerable complexity to economic analyses, effectively limit what might be achieved in formal frameworks, produce some paradoxes for expected utility theories, and qualify common inferences from conventional approaches. Besides these significant efforts of economists to establish alternative treatments of uncertainty, probabilists and statisticians such as Koopman (1940a, b), Good (1950, 1962a, 1988), Dempster (1967, 1968), Smith (1961), Levi (1974) and Suppes (1974, 1976) have stimulated a wider exploration of innovative (perhaps, more realistic and sceptical) formulations for intuitive, logical, comparative or epistemic probabilities, their potential range within upper and lower bounds, qualitative judgements, and novel belief functions – often in relation to what is warranted by evidence. A valuable account is provided by Walley (1991).

Keynes (1921) concluded that knowledge for any determinate economic agent can be obtained either through direct acquaintance (sensation, perception, understanding and experience) or through arguments about sets of propositions. He held that probability applies to the latter form of indirect knowledge, delineating various degrees of rational belief in propositions (rather than bound up with events) and associated with logical connections, as later extended and refined by Carnap (1950). The evidential basis of this Keynesian perspective shares three important difficulties with other approaches to epistemic probability (Walley, 1991: 23). First, the way in which probabilities are constructed from evidence needs careful elaboration and it must affect their interpretation and precision. Second, their subsequent use for inference and in descriptive statistics needs to be clarified. Third, we may also need a strong behavioural element to enhance understanding of the interactions among probabilities, evidence, current conduct and future inquiry (Walley, 1991: 109).

The continuity of Keynes' attachment to this logical version of probability – from his own first efforts and his appreciation of Ramsey's subjective theory to the treatment of uncertainty in *The General Theory of Employment, Interest and Money* (1936) and later Keynesian economics – is discussed by Bateman (1987, 1990) and O'Donnell (1990). Irrespective of whether Keynes recanted in favour of Ramsey's approach or persisted with a commitment to the logical theory for his macroeconomic theories, it is clear that he still maintained a very strong opposition to excessive presumptions of measurability and comparability, which was also supplemented by concerns over the application of statistical methods (regression and correlation) to establish simple and stable characterizations for the primary macroeconomic relationships.

Keynes considered that some probabilities are not comparable while other probabilities are not numerically measurable and non-additive. These limitations remove numerical probabilities from being the effective guides for the conduct of economic agents, when probabilities are summarized by mathematical expectations and other familiar statistics which demand substantial (costless) evidence to determine their values.

If ... the question of right action is under all circumstances a determinant problem, it must be in virtue of an intuitive judgement directed to the situation as a whole, and not in virtue of an arithmetical deduction derived from a series of separate judgements directed to individual alternatives each treated in isolation.

(Keynes, 1921: 345)

This rejection of basic mechanical calculations for characterizing actual choices and actions pervades *The General Theory*, in which most businessmen are assumed to be influenced by rational beliefs, intuition, general confidence, speculative anticipations about the 'psychology' of markets, personal circumstances, simple conventions and various imprecisions. Keynes (1921: 88) held that people are disposed to think of most probabilities as essentially connected to inductions of experience, causality and uniformity but, somehow, their historical treatment produced other bases as they fell 'into the hands of the mathematicians'.

Hicks and Allais also struggled to reconcile the character of probabilities, their measurability and potential neglect with the actions of economic agents. Allais, for example, recognized that some probabilities with smaller values are frequently ignored and that many calculations underlying choices avoid the comprehensive detail involved in mathematical expectations, while disputing the subjective approach of Savage and other EU-theorists. The views of Hicks are more difficult to express because they changed during his professional career. Initially, as he developed the distinctive portfolio approach to money and financial assets (Hicks, 1931) over two decades, he used means and variances to fully characterize the attributes of alternative portfolios and shared risks. Gradually, at mid-career and without much concern for measurability, he came to recognize a need to use

117

higher moments because of the evident non-normality and asymmetry among potential returns to financial assets.

Finally, a mature Hicks (1979) questioned most economists' treatment of causality, abandoned frequentist notions which were based on replication, and assumed the truncation of probability comparisons to those which can occur in a practical and incomplete range. Over fifty years, he shifted from an early reliance on theoretical averages over measurable risks to more realistic images of incompleteness and non-stationarity, perhaps because of his long interest in monetary theory which he felt was 'less abstract than most economic theory' and more often centred on historical episodes (Hicks, 1967: 156). Citing Keynes and Jeffreys, Hicks also insisted that stable numerical frequencies are insufficiently wide for the economic context.

> The probabilities of 'state of the world' . . . formally used by economists, as for instance in portfolio theory, cannot be interpreted in terms of random experiments. Probability, in economics, must mean something wider.
>
> (1979: 107)

Often incomplete information may only be sufficient for comparative judgements or it might support neither numerical probabilities nor comparisons. There is, therefore, a small hierarchy of probability theories for economists to consider in the light of informational quality and non-experimental conditions – with the various levels partially characterized by weak measurability and its consequences. Hicks' views seem unaffected by any practical assertions that 'in ordinary experience expectation rather than probability is the more widely used concept' (comment by Suppes on Fishburn, 1986: 349) or by axiomatic treatments with expectation as their principal element (Whittle, 1970).

Within our mathematical literature, unsure probabilities with interval-based or other formal representations have been explored by Good, Smith, Dempster, Koopman, Levi and others in a wide range of situations, including ones which continue to attract the use of 'precise-looking formulae' and statistical interpretations. Their efforts yield some awkward compromises or qualitative middle paths between polar attitudes, two of which are illustrated by Suppes and Good:

> to insist that we assign sharp probability values to all of our beliefs is a mistake and a kind of . . . intellectual imperialism . . . .
> On the other hand, a strong tendency exists on the part of

practising statisticians to narrow excessively the domain of statistical inference, and to end up with the view that making a sound statistical inference is so difficult that only rarely can we do so, and usually only in the most carefully designed and controlled experiment.

(Suppes, 1976: 447)

The approach is often only semiquantitative because of the difficulty or impossibility of assigning precise numbers to the probabilities. Some people will argue that it is misleading to use precise-looking formulae for concepts that are not precise, but I think it is more leading than misleading because a formula encapsulates many words and provides a goal that one can strive toward by sharpening one's judgments. Also it is easier to make applications to statistics if one has a formula.

(Good, 1988: 386)

A brief review of one compromise is offered by Good (1992), while the various sets of axioms for unsure probabilities are collected and assessed by Fine (1973), Fishburn (1986) and Walley (1991).

Against the unqualified acceptance of practical compromises, we might insist that 'a theory of rational belief needs to respect the limited precision with which we can scale our confidence in our beliefs' and that 'ignorance should be given its due' (Fine, in his comment on Fishburn (1986: 353), which cites alternative approaches such as envelopes and undominated lower values). How imprecision and ignorance are recognized is essentially a dynamic issue, with iterations among black-box mathematical formulations, discernments and beliefs or judgements (Good, 1962a: 321) determining outcomes of research efforts, choices and decisions, and relying on a host of supplemental numerical indicators and informal suggestions.

In practice, it might be appropriate to replace any second and higher-order probabilities (e.g. estimated standard errors for estimated coefficients in regression analyses) by alternative representations of uncertainty that are more explicitly vague, and to move from the basic rationality and logical consistency that has been characterized by familiar probability axioms and EU-theories to a 'rationality of type 2' (Good 1962b, 1971, 1988) and other notions, which recognize the costs of theorizing and calculating, complexity, the urgency of making decisions and other sensible qualifications to our abstract concepts that will stem from actual circumstances, their evolution, and the inherent consequences for probabilities from both flawed

119

reasoning and intervention of new information – all practical compromises in other guises.

The normal occurrence of inconsistencies or contradictions will partially depend on costs of thinking and calculating. Good states his type 2 principle of rationality as a recommendation that some allowance be made for such costs in attempting to apply EU-principles (or type 1 rationality). Thinking, by affecting mental states, must connect dynamic probabilities to partially ordered subjective probabilities, and major obstacles to effective thinking – such as a finite time for introspection, limited intelligence, gross errors and blunders, even insanity (Banks's comment on Good, 1988: 404) – can generate a series of 'type n' rationalities which influence the development of imperfect priors and call for both sensitivity analysis and a better clarification of temporal coherence. Clearly, it is also possible to extend the hierarchy of new rationalities to artificial intelligence (Good, 1988: 412).

WEIGHT OF EVIDENCE

A major source of difficulty for clarifying and interpreting probabilities is the inconsistency of evidence whenever they are based at least in part on evidence. As noted by Keynes (1921: 80), 'the weighing of the amount of evidence is quite a separate process from the balancing of the evidence for and against'. Information and degrees of confirmation are clearly involved in the latter and should not be confused with evidence itself when available evidence is inconsistent. Instead, we might wish to introduce a different notion of 'weight of evidence' with which to express degree of information or relative confidence, a notion that is distinct from probability. 'Weight' could, for example, be represented by a sum of favourable and unfavourable evidence, while probability is summarized by the difference between them. However, this ambiguous language implies that most weight calculations will be quite imprecise and perhaps subject to qualitative restrictions and approximation! Turing even suggested a unit of measurement (bans, decibans and natural bans) in terms of the smallest weight of evidence perceptible to the human mind.

Arguing provocatively that 'whole areas of statistics can be regarded as having their logical roots in the concept of weight of evidence and its mathematical expectation' and drawing on the EU-experience, Good suggests:

Weight of evidence can be regarded as a quasi-utility or epistemic utility, that is, as a substitute for utility when the actual utilities are difficult to calculate. (A quasi-utility can be defined as an additive epistemic utility.) Just as for money, diminishing returns eventually set in .... But the effect of diminishing returns can often be ignored. When this is done we naturally bring in the concept of expected weight of evidence.

(Good, 1988: 390)

Much more modestly and with logical probabilities specifically in mind, Keynes identified a simpler role of the weight for evidence or arguments:

As the relevant evidence at our disposal increases, the magnitude of the probability . . . may decrease or increase, according as new knowledge strengthens the unfavourable or the unfavourable evidence; but something seems to have increased in any case.

(Keynes, 1921: 77)

In choosing between two arguments with similar probabilities, we may just opt for an argument with greater weight. Unfortunately, this notion has never been properly clarified. Seidenfeld, in a comment on Good (1985: 264), suggests that if we treat weight of evidence as a means of determining whether an economic agent has sufficient evidence for a choice – as an indicator in a stopping rule which determines 'when it is no longer worth while to spend trouble, before acting, in the acquisition of further information' (Keynes, 1921: 76) – the groping procedure is seriously flawed and we remain unsure of what to do with the weight of evidence, not far from Keynes' own honest appraisal, even though we surely need some means of dealing with the accumulation of inconsistent information and with the truncation of decision processes. Good (1988: 411) seems more sanguine about stopping rules, whether driven by weight of evidence or by other criteria. He suggests that the process of stopping thinking and calculating (in accordance with his type 2 rationality) is 'what statisticians have usually done implicitly, to some approximation, for the last 200 years' and presumably he might extend this perspective to some economists, their clients and economic agents.

The widest assessment of this weighing notion is provided by Good (1985, 1988) and the earlier references that he cites. Good uses the logarithm of a 'Bayes factor' as a convenient numerical measure for the weight of evidence, describes the history of some related measures, illustrates situations where the terminology of Bayes factors and weights of evidence have more intuitive appeal, and covers a wide range of frequentist, subjectivist and mixed approaches to inference and statistical decision-making, all of which require additional supplements or sensible compromises to be effective.

DYNAMIC BELIEFS AND THEIR REVISION

Choice among alternative propositions often depends on the relative degrees of support for them which comes from incomplete or flawed evidence, with the quality and interpretation of 'support' or confirmation being expressed in a variety of different ways and not identified with numerical probabilities or presented within the constraints of probability axioms. A new logic of comparative support was launched by Koopman (1940a, b) for what he described as 'intuitive' probabilities. Hacking (1965) provides a brief sketch of this approach and a modest extension, while Fine (1973, chs. 2 and 7) gives a wider summary of various axiomatic formulations for comparisons which involve a simple notion that one experimental proposition (for Koopman, a statement about an event whose truth or falsity is determinable from the performance of an experiment) is at least as 'probable' as another proposition. Suppose H1 and H2 are two hypotheses, while E1 and E2 represent expressions of the corresponding evidence. Then we may be able to compare conditional propositions, e.g. H1/E1 and H2/E2, with the relationship (Fine: 183) which asserts 'H1 on E1 is no more probable than H2 on E2'.

Koopman left the direct selection of this comparative relation to individual intuition and gave no clear guiding principle for the ordering. Other logical theories, such as the attempt by Carnap (Schilpp, 1963; Carnap and Jeffrey, 1971) to develop a quantitative measure of support implied for a hypothesis H from evidence E with a confirmation function C(H/E), were more ambitious in advocating rational objectives. Carnap, for example, sought some intuitive or *a priori* bases to support the adoption of particular axioms which might narrow the potential class of confirmation functions to yield a single 'best' function suitably expressing support. However, the status of Carnap's concept for any rational decision-making remains unclear (Fine, 1973: 201), not least in justifying any acceptance of its arbitrary components for the economic context.

Recently, interest in probable reasoning has shifted to a new perspective of statistical inference, evidence and support which was

122

developed by Shafer (1976a, b) from a reinterpretation of the lower-bound epistemic probabilities used by Dempster (1967, 1968) and from an additional stress of a rule for combining the degrees of belief based on different bodies of evidence. Although Dempster maintained his personal commitment to practical Bayesian inference for statistics, the new theory for upgrading and combining vague beliefs in under-informed conditions, where familiar references to fair bets seem excessive, is usually known as the Dempster–Shafer theory of belief functions. This theory deviates considerably from its Bayesian roots, but has kept a concept of rationality in choices of intuitively attractive rules (e.g. those dealing with combination, conditioning, extension, conditional embedding and discounting) which are assumed to govern any operations applied to degrees of belief.

Numerical degrees of belief that emerge from the new approach are quite different from numerical chance (associated with aleatory experiments and unknown features of the world in which we live) and they are expected to obey different sets of rules. Further, there was no presumption of an objective relation between given evidence and a proposition which provides a numerical degree of support.

> I merely suppose that an individual can make a judgment. Having surveyed the sometimes vague and sometimes confused perception and understanding that constitutes a given body of evidence, he can announce a number that represents the degree to which he judges that evidence to support a given proposition and, hence, the degree of belief he wishes to accord the proposition.
>
> (Shafer, 1976a: 20)

This approach to degrees of belief is fundamentally different from both the objective concept associated with Jeffreys and Keynes and the subjective version promoted by de Finetti and Ramsey. It has a modest similarity to earlier ideas found in the discussions of non-additive 'probabilities' by Lambert and Bernoulli in the eighteenth century (Shafer, 1976b: 430; Shafer, 1978).

The drift within Dempster–Shafer theory from Bayesian notions and expected-utility axioms towards the constructive aspects of any 'small world' preferences can be traced through Shafer (1981, 1982b, 1986). The theory draws on some psychological theories (which we discussed in the section on heuristics and biases of chapter 5), mental experimentation, the weighing and decomposition of evidence, framing of epistemic probabilities and discernment, and individual images of evolving beliefs as new probabilities are constructed. Shafer and Tversky suggest:

> Probability judgment is a process of construction .... [It] depends not just on the evidence on which it is based, but also on the process of exploring that evidence. The act of designing a probability analysis usually involves reflection about what evidence is available and a sharpening of our definition of that evidence. The probability judgments we make may depend on just what examples we sampled from our memory or other records, or just what details we happen to focus on as we examine the possibility of various scenarios.

(1988: 263)

This perspective also admits the embarrassing possibility of dissonant and unreliable evidence from external sources, widespread discounting of weak information, problematical inference, and even 'wrong-headed' analysis – e.g. as associated with the Lindley paradox (Shafer, 1982b). A brief comparison with Shackle's concept of potential surprise is offered by Shafer, who indicates:

> The occurrence of outright conflict in our evidence should and does discomfort us; it prompts us to re-examine both our evidence and the assumptions that underlie our frame of discernment with a view to removing that dissonance. But the effort does not always bear fruit – at least quickly. And using all the evidence often means using evidence that is embarrassingly conflicting.

(Shafer, 1976a: 225)

The firmer recognition of mental processes, potential conflict and the process of construction may lead to a more sensible balance of judgement with incomplete or flawed evidence than is normally found in either Bayesian or frequentist analyses (Dempster, 1988). This recognition also spills over to many client–consultant interactions where the soft aspects of statistical analyses need communication, small-world omissions and sensitivity have to be clarified, and the design or data-collection protocols revised. The intrusion of such contextual elements leads to a liberalization (partial abandonment) of Bayesian frameworks and less decisive choices, which acknowledge the hazards of over-specifying the ability of client or consultant to produce useful probabilities and to follow prescriptive rules for suitable actions or decisions.

Given the substantial imprecision envisaged by Shafer, it is not surprising that this approach attracted attention in treatments of artificial intelligence and other major areas involving fuzzy concepts (Wierzchon, 1982), with the Sugeno measure being linked to Shafer's belief functions. A terse behavioural interpretation and critical assessment of the reasonableness of applying the Dempster–Shafer theory to various statistical problems is given by Walley (1991, sect. 5.13) within a useful survey of alternative approaches to imprecision, while different reactions to the theory are clearly illustrated by comments following Shafer (1982, 1986) and Dempster (1988). The implications of incomplete and flawed evidence and of construction processes for elucidation of subjective probabilities also affect the interpretation of difficulties identified in the final two sections below.

Yet another approach to evidence, rational changes of belief and the epistemic reliability of probabilities is due to Gardenfors and Sahlin (1982) and Gardenfors (1988), who focus attention on how minds are changed when new information arrives and is inconsistent with a previous epistemic state, 'an actual or a possible cognitive state of some individual at some point of time'. They distinguish among various mental exercises involving revision, expansion and contraction – such notions being linked to some rearrangements and speculative assessments of contingent changes in beliefs associated with retention or abandonment as consequences of new information are explored. Unlike Shafer and the proponents of heuristic models, Gardenfors uses an epistemological concept, an idealization of some underlying psychological concept that is to be judged by reference to rationality criteria or regulatory ideals. Rational states of belief are in equilibrium relative to internal criticism and they need not be matched by actual psychological states due to frequent shocks (disturbances from new inputs) and the ineffectiveness of such criticism.

From this perspective, numerical probabilities might seem to be unrealistically detailed (Gardenfors, 1988: 40) and a class of (epistemologically possible or permissible) probability functions may be maintained rather than a single one. Since some functions may appear more 'reliable' to the individual than others in the sense that 'they are backed up by more evidence or information than other distributions' (Gardenfors, 1988: 42), a concept of reliability has to be developed in the same way that Keynes (1921) felt it necessary to deal with the 'weight' of evidence and Shafer sought a similar concept. Similarly too, various aspects of individual beliefs that are relevant to decision-making cannot be represented by probability functions.

Reformulations of this dynamic framework for major changes in beliefs are actively stimulated by a dispute over the alternative 'foundation' and 'coherence' bases for changes. Gardenfors opts for the latter because it involves lower computational burdens and argues that the former (which requires satisfactory justification of beliefs) conflicts with observed psychological behaviour and thus is implausible. A coherence basis is clarified by Alchourron *et al.* (1985), but alternative versions exist and additional principles (e.g. invoking conservatism and the costs of acquiring information) must be introduced to choose among them.

ELUCIDATION OF PROBABILITIES

Presuming existence of objective or subjective probabilities in some sense, our interest shifts to issues of their elucidation, the quantification of their potential values, and the feasibility of reconciling various probability assessments, combining opinions or combining probability distributions (Genest and Zidek, 1986). In contrast to some of the views cited above, excessive precision can stem from a common 'scientific' insistence (the Kelvin dogma) that firm quantification is always desirable and possible. One attitude to basic measurability and its usual facilitation of mathematical analysis, even in regard to personal probabilities, is illustrated by Lindley.

> The scientist's appreciation of the world ... is of a collection of quantities, of things that can be described by numbers. For a scientist to understand and manipulate things, he must measure them, or at least think of them as things that, indirectly or directly, might be measurable .... Without this quantification the scientist cannot proceed: with it, he has at his disposal the full force of logic and the mathematical argument.
>
> (1982a: 66)

However, the success of scientific efforts must still depend on the crucial issue as to 'how well this quantification encapsulates the situation under study' and, for most economic phenomena with human activities, quantification is effectively qualified by the inherent complexities of contextual elements and by the usual flaws of human participants in acquiring suitable information, sifting through its inconsistencies, and computing suitable indicators of probabilities to guide decisions and choices. Obviously, there is a major hazard that the attempts at quantification may *not* encapsulate the relevant

126

features of any economic situations under study, but rather the attempts will distort subsequent perceptions of those situations.

Some awareness of similar hazards, beyond economic phenomena, has generated attempts to reinterpret probability statements and the outcomes of statistical analysis. Despite such attempts, e.g. as illustrated by Goldstein (1981, 1983, 1986), it has been difficult for economists, statisticians and others to give up the familiar convenience of common probability frameworks with their potential for spurious precision, awkward generalizability and inferential distortions.

Probability statements may shift from being the quantitative expressions of specific individual knowledge to purely technical intermediary quantities, which encourage the translation of some generalized knowledge into precise statements. Also, we may justify research practices in terms of simple metaphors and comforting 'as if' assumptions, which ease concerns over the incompleteness of specifications and measurability:

> it is worth emphasizing that as soon as we make any selection from the huge complex of assumptions (i.e. models) available to us, we are entering into a kind of metaphor. All models are metaphors .... I am loathe to abandon completely specified models, not because I take seriously all the implicit probability statements contained within them, but simply because studying certain aspects of the outputs from such 'as if' specifications leads me to useful understandings.
>
> (Smith in discussion of Goldstein, 1981: 121)

In practice, as effectively argued by Hayek and Shackle over the last fifty years, many economic events are essentially unique for decision-makers. Probabilities then depend on both special circumstances and personal judgements, with little connection to any laws of large numbers, independent repetition or convergence through learning. Further, the elicited values may not be free from major reporting errors, while subjective probabilities may be incoherent. The severity of practical and theoretical problems for assessments of such probabilities – located in both elucidation and debiasing – are clearly recognized by prominent Bayesians. An illustration is provided by Lindley et al. (1979), who describe the use of alternative criteria to improve probability assessments, the potential reconciliation of any inconsistencies (in line with Savage's promotion of rationality through greater awareness), and the weighting of opinions, while accepting the normal existence of human fallibility and its evident consequences for the properties of personal probabilities.

Other treatments of probability assessments and elucidation stressing the nature of reliability or practical calibration are conveniently surveyed by Lichtenstein *et al.* (1977) and explored by Dawid (1982). In recent years, discussion in this area has sought to clarify means of modifying, enhancing or extending assessments and to promote the use of analogous 'expert systems' and artificial intelligence, which reduce the computational burdens of complexity (Cooper, 1989; Lindley, 1987; Shafer, 1987). However, the weak bases for quantification of many probabilities in economics and other social sciences persist and they attract attention to basic questions identified with alternative non-probabilistic approaches to uncertainty (such as fuzziness, possibility theory, kaleidic perspectives and belief systems) and with claims for an 'inevitability of probability' (Lindley, 1982b).

DIALOGUE AND CONSENSUS-SEEKING

The evolution of science has often been described in terms of the emergence of some level of cognitive consensus among groups of scientists and the establishment of 'certified knowledge' (Cole *et al.*, 1978). Indeed, for some observers, science is essentially consensus seeking. With uncertainty, the achievement of consensus requires either a persistent process of Bayesian dialogue or the cataloguing of statistical 'facts' based on some accepted means of processing relative frequencies, both involving procedural rules for confirming or rejecting alternative theoretical perspectives in the light of empirical evidence and other considerations. There is an informal sharing and appraisal of estimated probabilities and some concern for the plausible convergence of these probabilities across individual values.

Similarly, for decision-making within economic organizations, most important choices reflect the active participation of many individuals and achievement of a consensus affected by the sharing of disparate information among those individuals (including both preferences and probabilities), unclear environmental contexts, and within-group social dynamics. Thus, rationality for economic choices and decisions requires effective communication and recognition of other inter-personal elements, which receive insufficient attention in myopic theoretical accounts that typically rely on the notion of representative individuals and thus ignore major consequences of the multi-person context. Both stochastic microeconomic theories and their macroeconomic counterparts still fail to deal with the practical problems that stem from the elucidation of probabilities, aggregation of individual preferences, and internal communication – significant features which characterize actual choice or decision situations for 'firms' and 'industries', and which substantially affect outcomes as acknowledged in recent accounts of informational flows in Japanese corporations.

Abstractions that ignore the costs of acquiring and processing information or the evident benefits of devising effective means of communication across the raw material suppliers, manufacturers and distributors (and 'horizontal' flows within different units of each firm) are increasingly challenged by newer 'business' perspectives on efficiency and rationality for managing economic activity. Such perspectives – identified with productive flexibility and the use of microprocessor technology – have begun to transform behavioural controls and practices in many larger corporations, as linked to globalization, benchmarking, integrated manufacturing, just-in-time inventories and other influential elements that are dependent on more informational awareness, the speed of adjustment to changing circumstances, and strategic or dynamic responses to competitive pressures. The essential substance of this transformation flows from the practical elucidation of imprecise but useful information (perhaps summarized by probabilities or alternative indicators), aggregation in the process of decision-making, and dialogue.

Any process of Bayesian dialogue presumes the existence of relatively firm convictions for personal probabilities that can be changed as more data become available, through argument with others with different convictions, and through individual reflection on coherence, the weight of evidence and basic foundations for values. Unfortunately, such dialogue has fragile underpinnings while useful communication is always restrained by numerical imprecision and by intrusive social factors. Coherence is attractive to statisticians, but may not be present in practice when expert advice is only part of the business nexus. Goldstein illustrates his own position on volatile personal probabilities or beliefs and his use of coherence as a statistician, which seems representative:

> I have beliefs. I may not be able to describe 'the precise reasons' why I hold these beliefs. Even so, the rules implied by coherence provide logical guidance for my expression of those beliefs. My beliefs will change. I may not now be able to describe the precise

reasons how or why these changes will occur. Even so, just as with my beliefs, the rules implied by coherence provide logical guidance for my expression of belief as to how these beliefs will change.

(Goldstein, 1983: 819)

Communication and effective dialogue may require the explanation of clear bases for plausible beliefs as a means of persuasion and for the support of general (often non-technical) reasoning, where any individual limitations in anticipating and assessing of evidence are openly recognized. This requirement need not indicate a formal framework or the familiar collections of referential axioms and statistical procedures including coherence (Goldstein, 1981, 1986), perhaps because the number of probabilities required for actual choices and decisions is prohibitively large, qualitative data are difficult to express with sample-space notions and short lists of possible outcomes, or because the actual uncertainties are rarely quantified in actual situations.

Technical advice on means of reaching consensus through the interactive revision of probabilities, e.g. as surveyed by Genest and Zidek (1986) and French (1985), is often awkward to adopt and difficult to justify on the basis of potential profits, competitive edge or other economic considerations because of the necessary qualifications, the excessive abstraction by statistical experts, haphazard imprecision, and the difficulties of explaining technical aspects to decision-makers. One unhelpful approach is illustrated by the efforts of Bergin (1989) and McKelvey and Page (1986) to reveal conditions under which some dynamic or iterative processes for the exchange of information, the announcements of individual posterior distributions, and subsequent reconsideration lead to common values among a group of experts. This abstract approach is too far removed from normal contexts for decisions, ignoring substantial features which characterize most convergent decisions beyond the confines of expert groups and negligible transactions costs.

A similar criticism affects many approaches developed for expert groups and Bayesian dialogues and outlined in professional journals over the last two decades – e.g. illustrated by comments of DeGroot (1974), Bacharach (1979), Sebenius and Geanakoplos (1983), and Bayarri and DeGroot (1989) in the *Journal of the American Statistical Association*. Given the real significance of consensus-seeking and uncertain data for economic organizations, it is surprising how little has been achieved in providing feasible and sensible methods for combining and transforming probabilities across individuals to facilitate group choices and decisions. A more sanguine view is offered by Spiegelhalter (1987) for expert systems in medicine, where discordant views are combined to provide a framework for computer diagnoses.

Complications in finding convergent values become more intense in the presence of irreconcilable differences and any uncooperative failures of intent among experts. These complications often appear in legal proceedings, for example, where the familiar inferential statistics are invoked to clarify or substantiate evidence (Meier, 1986) and where our professional associations have been unable to offer effective guidelines to codify experts' testimony so as to support their credibility and integrity.

7

ECONOMETRICS

The roles of probability in economics are most pronounced in econometrics, where estimation, stability, bounds for uncertainty, statistical inference and the validation of economic theories with empirical evidence are of primary interest. Economists quickly assimilated the basic notions of correlation and regression at the beginning of this century, and they actively sought cyclical and other patterns in economic time series with the periodogram and its successors after Fourier analysis was extended to stochastically disturbed data. This assimilation was directly recognized when the Econometric Society was established by prominent economists at the end of 1930, with its influential journal *Econometrica* appearing a few years later. By the end of the 1940s, economists were ready to analyse new collections of simultaneous equations and dynamic time-series models, absorb the Neyman–Pearson framework for hypothesis testing, rely on linear stochastic processes and a few asymptotic properties of conventional estimators (including bias, consistency, efficiency and distributional convergence) within stationary and ergodic contexts with little concern for historical conditioning or special circumstances, and accept the informational requirements of maximum likelihood approaches.

In later decades, they succeeded in widening the availability of economic data (time series and cross sections) and provided the easy software necessary to utilize a rapid growth of computational facilities. Recently, they have even begun to deal with measurement errors (again) and latent variables, non-stationarity, causality, asymmetric and chaotic time series, spectral methods, expectational difference equations, and various forms of statistical robustness in the presence of incomplete specification – flirting too with a partial replacement of estimation by the calibration of structural parameters. Thus, over six decades, economists have demonstrated an openness to the wider use of probability elements in empirical research and in discussions of econometric method. However, many econometricians persistently failed to show sufficient interest in the foundations of probability notions and the implications of various concepts which they adopted – despite an initial stimulus offered by Edgeworth, Keynes and Ramsey at the beginning of the present century, and later 'revolutionary' stimulus from prominent researchers at the Cowles Commission following the publication of Haavelmo's supplement on 'The probability approach to econometrics' to *Econometrica* in 1944 (Morgan, 1990a).

The search for probability foundations by Swamy *et al.* (1985), the promotion of more Bayesian analyses by Zellner (1988a), and the brief appraisal of a potential consensus across competing statistical paradigms by Durbin (1988) illustrate that some interest in such foundations persists. However, these illustrations are remote from the normal imperatives which drive applied econometric research and are expressed in most accounts of statistical methods in economics. Econometric textbooks reveal a pronounced lack of concern for the foundations of probability in regard to economic phenomena, while focusing on myopic accounts of estimation and inference in some well-specified abstract models.

The principal interests of theoretical econometricians are illustrated by contributions to the *Handbook of Econometrics* edited by Griliches and Intriligator, various econometric entries in *The New Palgrave* (Eatwell *et al.*, 1990b), the invited papers at World Congresses of the Econometric Society (Hildebrand, 1982), and some collections of readings such as Granger (1990), as well as various surveys contained in *Econometric Reviews* and the *Annals of Applied Econometrics*. These interests cover a wide range of areas involving the active use of probability by economists – e.g. see Leamer (1983a) on model searches and Hendry *et al.* (1984) on dynamic specification in the *Handbook* – however, an appropriate treatment of probability by econometricians remains elusive.

In order to clarify this elusiveness and describe the related activities of econometricians, our discussion begins with two short comments on some statistical conventions and aspects of exploratory or sequential research, before focusing attention on the so-called 'probability approach' to econometrics and on radical reappraisals which have changed both the character of empirical research and the communication or interpretation of findings during the last decade or so. A brief account of non-structural approaches, causality and the impact of long-term relationships is provided before we offer a few concluding remarks. The overwhelming impression of current econometrics is a bewildering mixture of firm assertions and weak theoretical under-pinnings, probabilistic language at odds with the actual research procedures, frequentist notions persistently misapplied to unstable economic environments and their subjective counterparts attributed to composite markets and other aggregates, confusion between causality and correlation, the neglect of basic measurement errors despite flawed data, excessive claims for the relevance of statistically significant findings and their potential generalizability, a haphazard reliance on economic theory and its weak linkages to data and particular circumstances, the promotion of statistical diagnostics for resolving disputes among alternative models despite the incommen-surability of these models and weakness of much empirical evidence, general failures to require sufficient replication of research findings (Anderson and Dewald, 1994; Lovell and Selover, 1994), and convic-tion that econometrics has a major role to play in the progressive development of our understanding of economic phenomena.

STATISTICAL CONVENTIONS

The statistical bases of econometrics are relatively simple. For the most part, they only involve the conventional frameworks for multiple regression, linear time-series models and stochastic processes which exhibit normality, stationarity and ergodicity to the extent that the common use of these frameworks is facilitated by computational software. More complex forms permit a wider class of non-normal distributions, limited ranges for economic variables, selection con-straints and Box–Cox transformations for sample data, stochastic and deterministic trends, and the combination of several relationships. Parametric estimation (usually based on arbitrary principles of max-imum likelihood, least squares and generalizations of least squares) is preoccupied with unbiasedness, efficiency and their asymptotic counterparts – assuming presence of sufficient stability, homoscedas-ticity and a modest amount of autocorrelation. Inferential procedures reflect an ambiguous attitude to the issue of whether models are taken to be true (and parametric estimates as evidence) or convenient approximations. As observed by Durbin in his presidential address to the Royal Statistical Society:

Statements about parameter values have been discussed as if parameters have a clearly-defined tangible existence, whereas in most cases they are at best mathematical artefacts introduced only in order to provide the most useful approximation available to the behaviour of the underlying reality.

(Durbin, 1987: 178)

Any greater recognition of temporary or ephemeral approximations and the inherent problems of relying solely on 'extreme hypotheses' (Simon, 1977) creates an awkward gap between the dominant rhetoric of statistical inference (and the declared properties of familiar estimates) and the modern data analyses outside the mainstream of econometrics, which often focus on robustness, inferential and parametric sensitivity, and the feasibility of useful statements about real situations, as illustrated in the topics raised by Tukey (1971, 1980).

Although relational specifications may be discarded because such estimates have undesirable values, model choice is primarily driven by the performance of significance tests or the outcome of simulation exercises. All of these ingredients are discussed with confirmatory language for the presumed models, even though empirical research is sequential and exploratory, models being adjusted as research proceeds without recognizing the important consequences of the adjustments for basic probability statements. Some researchers, however, note the hazards of unrecognized pre-test biases and the need to recognize the non-nestedness of successive hypotheses for parametric values and choices of variables (MacKinnon, 1983).

This image of econometrics is excessively simple, but it does not give a false impression of common behaviour in relation to probability. Bayesian approaches, perhaps more important in the future, have been handicapped by computational difficulties, higher costs relative to the numerical convenience of applying frequentist alternatives, and resistance to the strong specification of prior distributions for parameters. For tests of significance and maximum likelihood estimation, the comprehensive assumption of normally-distributed variables seems ubiquitous, even to the extent of being unstated in many articles and research reports! The corresponding presumption of normally-distributed parametric estimates (perhaps, as an approximation justified by distributional convergence and the application of laws of large numbers to stationary processes) almost always understates or ignores its weak theoretical support – given the non-reproduction of normality under non-linear operations, major

135

discontinuities due to prior choices based on probability cutoffs, non-normal measurement errors affecting economic variables, and the smallness of most samples available to economists. The tractability of normality generally overcomes a sharper perception of the actual characteristics of any research context.

Most conventional testing procedures reflect the framework developed by Neyman and Pearson (1933) and later outlined by Neyman (1952). This framework builds on four concepts for the asymmetric development of efficient and powerful statistical tests which guide the validation of hypotheses rather than discovery:

1 the existence of two kinds of errors possible to commit while testing a hypothesis;
2 the notion that these two kinds of error may be of unequal practical importance;
3 that a desirable method of testing hypotheses must ensure an acceptably low probability, say α, of the more important error; and
4 that point 3 being satisfied with an acceptable α, the probability of the less important error should be minimized.

It should be clear that there are few guidelines to assist in the wise choice of any probabilities for type 1 and type 2 errors in relation to potential sample size within this framework. Lehmann (1959: 61) and Good (1981: 155) have indicated an arbitrariness to statisticians, but economists seldom make the necessary connections between these three elements (Rosenkrantz, 1977); instead, they adopt the conventional 5 and 10 per cent boundaries for type 1 errors, or else use p-values (Gibbons and Pratt, 1975), when considering the acceptability of hypotheses. Clearly, such myopia is not restricted to economists. The impact of sample size is rarely understood, the common hazards of pre-selected significance levels are ignored, and statistical significance is persistently confused with substantive significance.

Suitable warnings for applied social sciences are provided by McCloskey (1985a, b), Guttman (1977, 1985), Hall and Selinger (1986) and others, but widespread ignorance persists. More generally, Cox (1977, 1986) outlines the nature of significance tests and their over-emphasis despite a remoteness of theoretical presumptions from actual research conditions. Dramatic reactions are exhibited in the promotion of iterative 'criticism', adequate robustness, and better samples (rather than optimal procedures) by Box (1979, 1980) to guide the evolution of specification searches, tentative models and practical diagnoses – primary objectives which rely less on statistical tests and their conventional interpretation, but are much closer to the 'hazardous undertaking of finding out what the world is really like' (Box, 1979: 1).

While the application of statistical techniques to economic phenomena is now commonplace, their acceptability was less certain in the late 1930s when Tinbergen launched a substantial expansion of their use to include economy-wide systems of equations, modern time-series analyses were in their infancy, and Haavelmo began to clarify a 'probability approach' to econometrics. Apart from their novelty, the principal problems with statistical techniques stemmed from a clear need to justify their value to a dubious audience, which questioned the consistency of probabilistic assumptions with the economic environment. The sharp exchange between a sceptical Keynes and an optimistic Tinbergen centred on their disagreement over whether this environment contained sufficient stability and whether statistical procedures could provide sensible values for any meaningful parametrizations of economic behaviour. These twin concerns are still important, but their visibility has remained dim since the rapid assimilation of probability-based notions, which emerged from the pursuit of Haavelmo's new programme at the Cowles Commission – despite an acrimonious confrontation with views found at the National Bureau of Economic Research over what constitutes 'measurement without theory' (Koopmans, 1947).

This programme began with a simple account of difficulties in reconciling the economic theories for business cycles and observed cycles, but it shifted to a more strident advocacy of statistical testing and maximum likelihood estimation (Haavelmo, 1940, 1943a, b; Koopmans, 1941), and full-blown promotion of complete probability distributions as the fundamental starting point for any empirical research (Haavelmo, 1944). The effectiveness of such promotion made this approach a central statistical convention for at least thirty years, despite the persistent efforts by critics to retain least square estimates (Wold, 1953; Wold and Faxer, 1957), weaken the common presumption of simultaneity by identifying recursive sequences within economic theory (Wold, 1964; Strotz and Wold, 1960), question the adequacy of sampling intervals (Bentzel and Wold, 1946), and recognize the latency of many economic influences which imply a need for 'soft modelling' (Wold, 1991; Joreskog and Wold, 1982).

SEQUENTIAL AND EXPLORATORY RESEARCH

For the last quarter century, the availability of adequate computer software has been sufficiently widespread to support a flood of empirical research which focused on regression equations, parametric estimation, tests of statistical significance and simple numerical diagnostics (such as the Durbin–Watson statistic and the coefficient of multiple determination). More recently, after some strong concerns were expressed with respect to the obvious hazards of 'data mining' and ill-directed research (as encouraged by easy access to the software, reporting inadequacies, and few attempts at replication), there emerged a strong interest in finding research procedures to generate a cumulative and progressive body of major research findings (Box, 1979; Hendry, 1980). However, this new direct interest in 'progress' did not lessen the general neglect of major factors associated with sequential and exploratory procedures which retain the language, formulae and confidence of one-time confirmatory frameworks.

The primary difficulties for statistical inference, e.g. based on t-statistics, stemmed from a general ignorance of power encouraged by the basic asymmetry of the Neyman–Pearson system for tests, and from a common failure to indicate changes to probability statements which are made necessary by the sequential character and incompleteness of exploratory research. These difficulties are compounded by flaws in economic data, due to measurement errors and awkward definitional issues, and by the remoteness of research from practical considerations. Bancroft (1944, 1964) noted the presence of pre-test biases and the conditional nature of weak specification, but their full impact was clear to few economists before the early 1970s. However, once the comprehensive accounts of Judge and Bock (1978) and Judge *et al.* (1980) appeared, econometricians should not have ignored them. Alternative and less firm research strategies, based on distributional robustness and a much wider collection of regression diagnostics, have been available since the end of the 1970s – as revealed, for example, by Atkinson (1986), Belsley *et al.* (1980), and Cook and Weisberg (1982). Moreover, many computational software packages were expanded to facilitate the adoption of these alternative strategies, the use of which serves to clarify the weaknesses of normal probabilistic foundations for any sequential statistical indicators.

The weaknesses of repeated significance tests can also be approached through technical discussions of the nesting/non-nesting of hypotheses or through wider issues of choice between general-to-specific strategies versus specific-to-general strategies. During the 1960s, step-wise computer programs were introduced to enlarge or diminish the size of regression equations in an automated way, with a few statistical indicators guiding choice between successive steps. It was realized that step-down procedures determined by the values of individual t-statistics receive partial support from the nesting of hypotheses in a research path. This support is clearly not present in the alternative step-up procedures, which seems to justify a tentative preference for general-to-specific strategies. Comparisons of rival models, however, cannot always be expressed in a simple nested format, so these strategies must be recast for wider choices, encouraging an active search for appropriate 'non-nested' statistical methods. Unfortunately, apparent success in this area can diminish the sensible use of non-statistical guides (e.g. economic theory, institutional constraints and any historical awareness) to choice at each step – since the sequential reliance on both statistical and other guides will produce a context of unclear probabilistic properties for any eventual choices. Beyond potential encouragement to the neglect of non-statistical considerations, the combination of multiple tests (simple t-statistics, autocorrelation indicators, checks for normality and heteroscedasticity, and other indices of model adequacy) must introduce the awkward spectre of 'path dependency', whereby the order in which tests are used will affect eventual outcomes of research procedures.

THE PROBABILITY APPROACH TO ECONOMETRICS

The important dispute between Keynes and Tinbergen over the potential applicability of statistical methods to economics was preceded by interpretative concerns for the regression equations used to represent either supply or demand functions. Given the persistent assumption of a moving equilibrium and the limitations of relying on static economic models, Moore (1908, 1917), Working (1927), Schultz (1938) and other agricultural economists (Fox, 1982, 1989) tried to indicate some major difficulties for empirical research and their partial resolution. These difficulties were associated with unmeasured theoretical variables, high degrees of intercorrelation among measured variables, inaccurate indices, the means of summarizing the relevant impacts of 'other' prices or past prices, and an inability to unscramble the simultaneous supply and demand influences with the data and statistical procedures which were available to researchers. Clearly, *all* of these elements must affect the feasibility and validity of inferences from any fitted equations, including those expressed in the language of probability and established within familiar statistical frameworks, so reliance on these inferences has to be effectively qualified.

The Keynes–Tinbergen dispute served to focus attention on such complications. Along with three other developments, it stimulated search for a better treatment of the probabilistic foundations of econometrics. These developments were:

1 the emergence of significance tests as identified with Neyman and Pearson;
2 an active promotion of maximum likelihood estimation along the lines established to guide experimental design in agriculture by Fisher, Snedecor and their followers; and
3 the rapid extension of a mathematical theory for asymptotic distributions, ergodicity and the large-sample properties of estimators for stable coefficients in linear difference equations and other contexts affected by bias.

This search yielded the 'probability approach' to econometrics and the emergence of identification, least-squares bias, structural autonomy, consistency, and statistical testing as primary concerns of econometricians, but it produced a myopic environment in which many of the important obstacles to applications of the new approach could be ignored. Haavelmo launched the new perspective with a strong assertion:

> It is sufficient to assume that the whole set of, say n, observations may be considered as one observation of n variables . . . following an n-dimensional joint probability law, the 'existence' of which may be purely hypothetical. Then, one can test hypotheses regarding this joint probability law, and draw inference as to its possible form, by means of one sample point (in n dimensions).
>
> (1944: iii)

This assertion was not justified by feasibility or sensitivity to the awkward features of any economic context. Rather, support for this 'more rigorous, probabilistic formulation of the problems of economic research' came from two other strong assertions:

First, if we want to apply statistical inference to testing the hypotheses of economic theory, it implies such a formulation of economic theories that they represent statistical hypotheses, i.e. statements ... regarding certain probability distributions. The belief that we can make use of statistical inference without this link can only be based on a lack of precision in formulating the problems. Second, ... there is no loss of generality in choosing such an approach.

(Haavelmo, 1944: iv)

The first of these justifications reflects a fundamental obstacle to statistical testing since economic theories cannot usually be represented by parametric constraints in well-defined statistical models without substantial distortion, excessive simplification, and a considerable amount of wishful thinking regarding the limited dimensions of these models, the compatibility of data intervals with those envisaged (or not made explicit) in economic theories, our ability to produce sensible approximations for any statistical specification, structural stability, and simple properties of model errors. (The recent reappraisals of econometrics, associated with Hendry and Leamer, for example, retain this desire for probabilistic foundations of hypothesis testing, but they recognize the weak bases for sensible specification and a need for much clearer procedures to guide estimation and inference beyond those recognized in the developments which triggered initial promotion of the probabilistic approach.)

The second justification is false in practice because severe losses of generality are introduced by the common dependence on tractable distributions, a few conventional statistical models, the arbitrary Neyman–Pearson approach to statistical tests (and its rivals) and asymptotic approximations. In any case, Haavelmo's proposals were amended as they came to be reflected in the later research of econometricians (Spanos, 1989), further weakening the clarity of probability foundations for what many econometricians actually do. This later divergence appears in any comparison of the high aspirations revealed in reports from the Cowles Commission by Koopmans (1950a) and Hood and Koopmans (1953), which developed the probability approach, with the more modest ones in subsequent textbooks and journal articles. Brief retrospective views on the later impacts of the Commission's efforts are given by Klein (1991) and Malinvaud (1991).

Three methodological aspects of the full probability approach are especially interesting:

141

- the prior classification of variables by reference to their statistical character (Koopmans, 1950b);
- the need for complete structural specification and the potential to identify individual constituent equations by zero constraints and prior information (Koopmans, 1949; Koopmans, 1950b); and
- the existence of sufficient structural autonomy and the nature of conditional predictions (Marschak, 1953; Aldrich, 1989).

Endogenous variables are distinguished from exogenous ones solely by their presumed non-independence of equation errors, while any predetermined variables are independent of contemporaneous errors. Given the coexistence of more than one endogenous variable in any structural equation, this choice for the separation of variables clearly indicates a preoccupation with asymptotic bias, perhaps to be eliminated by the use of instrumental variables or other novel methods of estimation which may be free from 'least-squares bias', although not from bias in finite samples. With this particular choice, Haavelmo's approach becomes more limited as there are no obvious grounds for accepting the primacy of asymptotic criteria when dealing with phenomena so unlikely to be stable over a long time, and as the prior use of statistical tests for exogeneity or predetermination (rather than relying on firm prior knowledge) adversely affects the derivation of statistical properties for estimated parameters.

All theorems for the identification or distinguishability of linear structural equations (by omissions of variables or equality constraints) require an accurate classification of the endogenous and predetermined variables within a statistically-complete system of equations. This is a strong requirement that cannot reasonably be expected to hold for suitable approximations to most economic contexts due to the general vagueness of economic theories and the intractability of dealing with a large number of equations (Orcutt, 1952). Indeed 'under-identification' might be perceived as common, but yet distorted, by an excessive truncation of equations, which gives the impression of widespread identification and supports the use of 'system-based' estimators, as argued by Liu (1955, 1960).

The convenience of completeness also extends to a derivation of alternative forms – e.g. structural, reduced and resolved forms – of a system specification and to the connections among the corresponding probability distribution functions. Without completeness or some type of optimistic conditioning/separability (perhaps linked to a causal ordering of variables) for groups of structural equations variables (Simon, 1953; Ando et al., 1963), the status of any probability approach seems unclear.

One concern behind adoption of the probability approach was the hope of finding a superior basis for economic predictions, as generated by the combination of empirical evidence and economic theory within a simple statistical framework. The usefulness of these predictions always depends on the intertemporal stability of economic behaviour and institutions, as well as the way in which this behaviour is represented by 'structural' approximations. An isolation of change within some part of a perceived structure can be discussed with the concept of 'autonomy', initially introduced by Frisch to explore potential invariance of structural parameters in a given model, or collection of structural equations, under the conceivable changes occurring elsewhere in the model. The search for appropriate means of estimating the structural parameters to generate conditional predictions, rather than relying on estimates for their reduced-form counterparts, is driven by the common view that structural equations are less susceptible to change (adversely affecting predictions) than are the reduced-form equations obtained from them because, if suitably chosen, the former will enjoy more autonomy relative to controlled changes.

The apparent success of the rational-expectations challenge to econometrics in the last two decades stems from a display that this old view of relative autonomy is false when anticipations and other factors produce cross-equation restrictions that reduce structural autonomy. In any case, the translation of the simultaneous-equation theory into large commercially-successful models (routinely selling predictions to governments and businesses) is associated with major 'tender loving care' and tuning adjustments, described by Howrey et al. (1974), which permit a much wider framework for predictions involving non-probabilistic and qualitative information that cannot be readily integrated in normal methods for parametric estimation.

Finally, the use of a probability approach need not rely on frequentist notions and a preference for one kind of statistical test. Although Haavelmo's views were strongly affected by the frequentist developments of the late 1930s and early 1940s, the later emergence of feasible Bayesian alternatives provided another challenge to the conventional reliance on significance tests and maximum likelihood when structural formulations are sufficiently small and the computational burden is relatively modest. Zellner (1971), Box and Tiao

(1973), Rottenberg (1971, 1973) and Marales (1971) illustrate the slow development and assimilation of these Bayesian alternatives until the beginning of the 1970s, while Zellner (1979 1988a, b) deals with subsequent developments in the promotion of a Bayesian perspective for econometrics, which is likely to become more visible with the rapid growth of computing capacity, especially graphing, over the next decade.

RADICAL REAPPRAISALS

Although the Cowles Commission's SEM approach quickly came to dominate econometric textbooks and many research reports during the 1960s, some primary characteristics were substantially modified and its probabilistic ideals were compromised as researchers failed to recognize the severity of identification notions, the absence of sufficient information to support distributional assumptions as the basis for maximum likelihood estimates, the severe consequences of exploratory research in a framework that was essentially framed as confirmatory, expectational linkages which may undermine structural autonomy, the potential for systemic instability, and the impact of temporal aggregation on linear stochastic processes. Despite the commercial success of economy-wide models and the large volume of econometric research, there seemed to be a lack of progress which encouraged fundamental challenges to the comfortable vision offered in the textbooks – radical reappraisals that promote changes in econometric method, discard comprehensive models for their smaller alternatives, demonstrate the widespread incidence of whimsical specification and general failures of robustness, seek Bayesian and system-theoretic perspectives, and give greater attention to time-series characteristics of economic phenomena. Some major features of these reappraisals are clarified by the surveys of Charemza and Deadman (1992), Darnell and Evans (1990), and Pagan (1987), while the dialogue of Hendry et al. (1990) demonstrates the absence of a new consensus on appropriate econometric procedures, interpretations and objectives. We illustrate the reappraisals by looking at some views of prominent contributors to recent debates among econometricians, and we draw attention to the corresponding probability issues.

The most successful reappraisal of econometrics is associated with Hendry and his associates, who provide a statistical framework for sequential testing and model specification involving concepts such as an underlying data generating process and encompassing of alternative economic theories. Choice in this framework is driven by a given collection of design criteria and basic probabilistic indicators, which generally result in procedures giving excessive weight to statistical factors – while overstating the validity of familiar test statistics; the ability of researchers to draw firm conclusions from empirical findings; and the overall feasibility of comparing rival economic models within some simple structures that are severely constrained by data availability, data quality and research conventions. These structures, distinct from the data generating function, are usually regression equations defined over aggregate data collected at uniform time intervals, and they presume stationary error (ARMA) processes defined over the same intervals.

> Observed economic data seem to be the (unique) outcome of a complicated, unknown, and perhaps unknowable, data generation process on which they provide some evidence, albeit needing careful interpretation. All empirical models thereof are ... highly abstract simplifications with properties which are a function of the underlying mechanism and are best deemed 'false' *only* if rejected against other models.
>
> (Hendry, 1983a: italics added)

Hendry (1980: 389) goes so far as to restrict econometric theory to simply 'the study of the properties of data generation processes, of techniques for analysing economic data, methods of estimating numerical magnitudes of parameters with unknown values, and procedures for testing economic hypotheses'. This view surely reduces such theory to a modest part of statistical theory or an addendum to it, far from the much richer vision offered when the Econometric Society was established, as revealed in the editorial by Frisch in the first edition of *Econometrica* (1933) – unless a qualifier 'applied' indicates all aspects of econometrics outside statistical theory (as concerned with estimation and inference)! The myopic view expressed by Hendry is also remote from the multi-dimensional vision offered by Samuelson *et al.* (1954) in their evaluation for the Society, which preserves the inclusion of economic theory and other forms of intellectual activity within econometrics.

Similarly, the view is at odds with the vision at the Cowles Commission when the current fashions in econometrics were being initiated:

> Econometrics is a branch of economics in which economic theory

and statistical methods are fused in the analysis of numerical and institutional data.

(Hood and Koopmans, 1953: xv)

Nor is this remoteness eliminated by putting 'data admissibility' and 'theory consistency' among the basic criteria for justifying a preference among alternative empirical models (Hendry, 1983b). If econometrics is a sensible amalgamation of economics, statistics, mathematics and computing (Frisch, 1933a), then econometric theory includes theoretical elements drawn from *all* of its constituent parts, without the radical asymmetry that gives statistical theory special and unjustified status – an asymmetry at odds with a long tradition of statistical analysis by economists (Tintner, 1968), extending from Cairns and Jevons to the present time.

Since the primacy of the Cowles Commission' SEM approach was established, the role of economic theory, institutional knowledge and data (rather than statistical theory) in specifying econometric relationships has been subject to diverse opinions. Koopmans (1945) insisted that some prior information on active variables was 'truly indispensable' for the identification of these relationships, while Klein (1982) indicated the meaningful use of a wider collection of economic theoretic restrictions. Earlier, he insisted:

The building of institutional reality into . . . economic relationships and the refinement of basic data collection have contributed much more to the improvement of empirical econometric models than have more elaborate methods of statistical inference.

(Klein, 1960: 867)

On the other hand, time-series analysts such as Granger suggest that adequate specification is much beyond the scope of macroeconomic theory:

at most it may indicate potentially relevant variables for inclusion in each equation. Frequently the theory is little more than a plausible story, and only rarely does the theory being applied help to specify the lag structure of the variables.

(Granger, 1981b: 123)

Against the backdrop of this diversity, Hendry's vision reduces the earlier focus on economic theory and institutions, but he is not alone in questioning the excessive claims for such theory in the pursuit of

146

empirical research. Unfortunately, he drifts towards the opposite extreme and exaggerates the potential contribution and rigour of statistical inference with proclamations that the 'three golden rules of econometrics are test, test and test' and that 'rigorously tested models, which adequately described the available data, encompassed previous findings and were derived from well based theories would greatly enhance any claim to be scientific' (Hendry, 1980: 403). This wishful thinking should be compared with the disappointment expressed by Haavelmo (1958: 355), who blames the weakness of concrete results from econometric research on 'the shortcomings of basic economic theory' and 'the somewhat passive attitude of many econometricians when it comes to the choice of axioms and economic content of the models we work on' rather than on the need to extend our statistical procedures.

The new concept of encompassing is attractive at first sight because it seems compatible with the desire for 'progress' in our understanding of economic phenomena, as encouraged over time by the cumulative evaluation of empirical findings. Descriptions and brief applications of the concept are provided by Hendry (1988), Mizon (1984) and Mizon and Richard (1986). The clear danger and impracticality of taking the concept too seriously are evident in the excessive assertion by Ericsson and Hendry:

> it is important to examine *all* available evidence on an empirical model *jointly* rather than simply corroborate a subset of the implications of a theory. Only well-tested theories that have successfully weathered tests outside the control of their proponents and can explain the gestalt of existing empirical evidence seem likely to provide a useful basis for applied economic analysis and policy. That means *encompassing* the evidence with a *congruent* empirical model.
>
> (1989: 13, italics in original)

More modestly, encompassing can be seen as a useful requirement that researchers try to explain the relevant empirical findings of rival models in the context of a model that they favour and with the data which persuaded them to support the latter.

Clearly, we cannot expect to combine all existing information on some given economic phenomena, and rival interpretations may be better expressed with diverse collections of data – often with different time intervals, measurements and statistical procedures. Many economic theories are incommensurable and incomplete, some theories cannot be sensibly displayed within a regression or time-series frame-work, and choice of a general (approximate) framework for compar-isons may prejudice conclusions. Problems with a full commitment to encompassing grow when the 'evidence' on any relative validity of theories refers, primarily, to test statistics based on erroneous formulae because of confusions, pre-test complications, excessive distributional assumptions, and a persistent reliance on asymptotic approximations. There are few tidy answers in empirical research, we are often unsure of the role that 'useful' inferences play in the formulation of policy advice, and theories are seldom rejected on the basis of empirical findings alone. Standard design criteria for the encompassing-testing approach and corresponding sources of information are summarized in Hendry (1983b: 199) and Ericsson *et al.* (1991: 10a). It is remarkable that their robustness and limitations are not made more explicit, especially since probability has such a major role in their interpreta-tion. For economists, the failures of the criteria to reflect special historical conditions, many forms of non-stationarity, relevant time intervals unmatched by data, aggregation issues, almost all institu-tional characteristics, and a host of other complications should eventually reduce the attractiveness of the approach and revive an earlier wider vision of the ultimate goal of research, including a more modest concept of evidence.

> Many kinds of work can contribute to this ultimate goal and are essential for its attainment: the collection of observations about the phenomena in question; the organization and arrangement of observations and the extraction of empirical generalizations from them; the development of improved methods of measuring or analysing observations; the formulation of partial or complete theories to integrate existing evidence.
>
> (Friedman, 1952: 237)

In another important reappraisal of econometrics, Leamer (1978, 1983a, b) stresses a tension between the precise assumptions of familiar statistical approaches to inference – focused on the targets of unbiasedness, consistency, efficiency, and statistical significance – and the haphazard methods actually used by many economists to analyse non-randomized data and test their theories after an ill-structured search across alternative specifications. The outcome of this tension is difficult to clarify using any basis in probability. Statistical inferences become fragile and need to be accompanied by some form of sensitivity analysis without a prior commitment to precise distributional assumptions, as recognized by Box in his use of 'criticism' and by Tukey and Mosteller in their promotion of robustness and exploration as a fundamental criterion guiding normal research activities. Leamer advocates a new Bayesian approach of interpretive search, which avoids sharp probabilities, recognizes approximations and a widening of options after the appearance of anomalies, and resists the common excesses of unqualified conclusions and whimsical inference in relation to economic phenomena.

> I believe that the Bayesian calculus applied to interval probabilities offers a proper foundation for inference, and the extent that our analyses do not conform with this norm, it is the analyses, not the norm, that ought to be adjusted.
>
> (Leamer, 1985a: 299)

Inevitably, this view attracts criticism, not least from those opponents to traditional methods of economic research who retained and re-emphasized frequentist notions and an excessive reliance on hypothesis testing and maximum-likelihood specifications. Criticism is encouraged by the colourful language adopted by Leamer (1983b, 1985b) and the painful accuracy of his assault on the 'sad and decidedly unscientific state of affairs we find ourselves in' which affects the communication and acceptance of quantitative evidence.

> Hardly anyone takes data analyses seriously. Or perhaps more accurately, hardly anyone takes anyone else's data analyses seriously. Like elaborately plumed birds who have long since lost the ability to procreate but not the desire, we preen and strut and display our t-values.
>
> (Leamer 1983b: 370)

This is a serious obstacle to progress in an empirical science – one that cannot be removed by more rigorous testing. When, for example, MacKinnon (1983: 107) notes 'applied econometricians are usually very sloppy about testing their models' and then argues 'anything that forces them to test models a bit more rigorously is *therefore* highly desirable' (emphasis added), he seriously misrepresents both the obstacle to better empirical research and its implications. Use of non-nested hypothesis tests to discriminate among alternative models and to confront a favoured model with evidence, recommended by MacKinnon, does not remove whimsical inference. His forceful suggestion that 'doing this must increase the probability that the model finally selected will not be thoroughly false' illustrates again the wishful thinking that underlies persistent advocacy by econometricians of myopic statistical tests.

Because of pre-test biases, non-normality, constraints on data intervals, measurement errors and a host of other relevant factors, we really have little idea of the true probability characteristics for fashionable statistical indicators, as derived from the limited samples and background information available to economists. Hendry (1974: 576) is surely correct when he insists 'it seems important to examine every aspect of a model as thoroughly as possible, since even if interminable revision is not practical, we should know what problems are untreated, rather than just hope that they are absent'. However, all current schemes for repetitive statistical testing with flawed probabilistic bases are insufficient as the focus for examination and appraisal of model specification (and derivation of sensible inference) in economic research, especially if we accept propositions that all models are essentially 'false' and existing data seldom refer to the fictional agents of economic theory. Indeed, such schemes seem so preoccupied with statistical calculations and the presumption of immediate conclusions that they almost lose sight of the normal economic environments, weaknesses of non-statistical information, inherent hazards of probabilistic metaphors and 'as if' data generating processes for econometrics (Leamer, 1987), and the instabilities of human-based relationships when presenting their ephemeral conclusions for some wide approval. Perhaps this substantial distance from the real world of economic phenomena explains why the severe judgement on econometrics by Leontief still remains valid, twenty-five years after it was expressed.

> In no other field of empirical inquiry has so massive and sophisticated a statistical machinery been used with such indifferent results. Nevertheless, theorists continue to turn out model after model and mathematical statisticians to devise complicated procedures one after another. Most are relegated to the stockpile without any practical application or after only a perfunctory demonstration exercise.
>
> (Leontief, 1971: 3)

If any probabilistic focus is to be maintained in econometrics, then the only effective counter-response to this judgement is a demonstration that we can produce a body of established 'applied' inferences on economic phenomena which justifies the adoption of this focus. The sustained arguments of Hendry and Leamer for new guiding princi-

150

ples and sequential procedures indicate that past efforts since the birth of the Econometric Society have failed to meet this standard, despite the considerable resources committed to statistical calculations.

In a third major reappraisal of econometrics, Kalman (1983: 98) insists that Haavelmo's aspirations to give a solid foundation by the 'dogmatic application' of probability theory have not been fulfilled because this theory says nothing about the 'underlying system-theoretic problems' which stem, at least in part, from the complicated nature of economics.

> There are no 'laws' in economics ... because economics is a system-determined science .... If we need a Keynes to tell us how to interpret the 'law' of supply and demand in our own environment, we could hardly claim from that 'law' coequality with the majesty of Ohm's law which every lowly electron must obey without the luxury of having Keynes as a consultant.
>
> (Kalman, 1980: 6)

> The non-universality or system-determinedness of economic insight is ... most inconvenient for the normal practice of modelling .... The only alternative is not to base economics on handed-down theory but to proceed directly *from data to model*. This frequently happens when the situation is too complicated to permit using a 'clean' theory.
>
> (Kalman, 1980: 6–7, italics in original)

This perspective is obviously difficult to reconcile with the usual presumption of stable parameters for econometric representations and with familiar tests of statistical significance which impose sharp constraints on these parameters. As Kalman notes:

> Under the influence of the standard statistical paradigm, econometricians have been assuming, unquestioningly and for a long time, that a system is the same as a bunch of parameters.
>
> (1983: 98)

Further, the current modelling procedures are seriously weakened by a host of prejudicial elements embedded in 'assumptions unrelated to data, independent of data, assumptions that cannot be verified or contradicted from data' (Kalman, 1983: 125), illustrated by the *ad hoc* separation of endogenous and exogenous variables, arbitrary specifications of potential correlation among equation errors and between such errors and explanatory variables, and the neglect of measurement inaccuracies. Other aspects of this reappraisal are addressed in Kalman (1982 a, b).

In a further reappraisal, Lawson uses the Keynes–Tinbergen dispute to focus attention on philosophical concerns that are often neglected in the debates among econometricians. In particular, he stresses the basic distinction between realism and instrumentalism (Lawson, 1989a; Bhasker, 1989), both of which are to be found in the reasoning and practices of past research. Realism involves an assertion that the objects of analysis exist independently of the enquiry in which they are objects, while instrumentalism involves the doctrine that some successful prediction (a theory consistent with a given set of empirical data) is all that is required. Here, Lawson suggests that 'most prominent critics of the "probability approach in econometrics" reveal an implicit realist position on relevant issues' and so too do many proponents and practitioners of this approach, but

> because of certain premises that must be accepted if econo-
> metrics is to be advanced, then, whatever their philosophical
> inclinations might otherwise be, the early proponents of the
> econometric approach – and the same is generally true of current
> day practitioners – appear to turn ultimately, to a form of
> instrumentalist reasoning.
>
> (Lawson, 1989a: 237)

This muddled situation leads to the persistent ambiguities that separate the usual contents of textbooks from applied econometric research, and continue to obscure dialogue on econometric method and the interpretation of empirical data. Wider aspects of this philosophical challenge are outlined in Lawson (1987a, b, 1988, 1989b), while Leplin (1984), Putnam (1987) and Churchland and Hooker (1985) illustrate other aspects of realism.

NON-STRUCTURAL MODELS, CO-INTEGRATION AND LONG-TERM RELATIONSHIPS

Much of econometrics using time-series data during the last fifty years has either presumed covariance stationary and ergodic properties for stochastic processes or has employed differencing and simple trans-formations to seek such properties. Both equation errors of structural regression models and time-series models are typically represented by a few autoregressive-moving average (ARMA) processes, with con-stant parameters which imply the fulfilment of these two properties.

Historical incidents and non-random shocks are identified with specific dummy variables, the particular values of explanatory factors that are explicitly included in the signal components of regression models, and interventions in time-series models. This conventional treatment reflects a long-established conflict between history and econometrics, clearly stated by Wiener who fostered the spectral approach to analyses of linear stationary time-series data and stressed optimal prediction defined in terms of stationary processes and mean square errors.

> A scientific theory bound to an origin in time, and not freed from it by some special mathematical technique, is a theory in which there is no legitimate inference from the past to the future.
>
> (Wiener, 1949: 26)

Covariance stationarity and ergodicity associated with invariance over time (found in probability-preserving and measure-preserving transformations) offer a comfortable and familiar framework for which most technical problems have been solved. Unfortunately, the evolution of economic phenomena may be essentially non-stationary, bound to historical time and not free from the vagaries of changing conditions, so any mathematical devices that are preoccupied with a false presumption of comprehensive invariance are a poor substitute for weaker devices that enable significant historical elements to be explicitly recognized. Surely technical convenience is not synonymous with empirical relevance, even if some modest degree of numerical predictability is sought? Legitimacy of predictions is not enhanced by ignoring the historical and non-stationary nature of economic phenomena and exaggerating the scope of convenient probabilistic simplifications.

The first intrusion of stationary processes into discourses of economists occurred when Yule (1926) sought to correct the excesses found in an earlier search for the causes of economic fluctuations. He pointed to the potential for some non-structural 'mimic cycles' to undermine the empirical support for theoretical perspectives. Yule (1926) and Slutsky (1927 [1937]) introduced autoregressive and moving average models, in which cyclical processes can emerge through the combination of linear transformations and a sequence of random shocks, later termed 'white noise' shocks – rather than through the distributed impact of primary economic causes, as expressed in structural models.

153

Development of this erratic-shock approach and newer theories of stochastic processes (essentially ARMA or ARIMA processes, until recently) provided both a non-structural or time-series approach to econometrics (Wold, 1954; Box and Jenkins, 1976; Morgan, 1990b) and the framework of optimal predictors, linear processes and rational expectations for economic theory and regulation (Whittle, 1963). If assumptions of linearity, normality and stability are added, the asymptotic properties of the corresponding parametric estimates can be appraised.

Once time-series concepts are extended to multiple equations, they:

1 give a limited means for discussing both causality and exogeneity, as indicated by Granger (1980, 1982, 1988) and Sims (1972a, b);
2 offer a platform for condemning the excesses found in discussions of the *a priori* restrictions for structural models, illustrated by Sargent (1979), Sims (1986) and Litterman (1984); and
3 attract the replacement of estimation by calibration with equations used to display the rival economic theories of business cycles rather than test them.

Unfortunately, the usual empirical treatments of causality and exogeneity rely on correlation, statistical tests and parametric constraints to provide recursive systems so a confusion with purely statistical characteristics is always possible and the fruitfulness of 'causality tests' is open to considerable doubt – not least in the absence or weakness of subject-matter content, associated with non-experimental data and economics, but also due to the awkward probabilistic elements for testing. A survey of newer causality treatments is provided by Vercelli (1992). Engle *et al.* (1983) illustrate how far the stochastic definitions of exogeneity have been extended from that earlier used at the Cowles Commission, while revealing few reasons for those definitions to be applied to real economic contexts or for them to be sufficiently insensitive to specification errors, transformations of variables and measurement errors. The primary criticisms of causality tests are summarized by Jacobs *et al.* (1979), Feige and Pearce (1979), and Conroy *et al.* (1984).

Although the excesses in structural specification are highly visible – but not sufficient to justify the premature *post mortem* on 'Keynesian macroeconometrics' by Lucas and Sargent (1979) – the need for much greater scepticism for interpretation of econometric models and their empirical findings is obvious. However, just being different from the conventional format of the Cowles Commission, especially by relying on different or fewer parametric constraints (e.g. drawn from the optimizing behaviour of microeconomic agents without much consideration of aggregation issues) is not sufficient justification for the adoption of any non-structural time-series model. Flaws in one methodology do not inevitably lead to support for any one alternative methodology, so the use of non-structural models needs to be sustained by other means. It is not clear that probabilistic elements have much to do with such means, even when a qualifier 'Bayesian' is attached to some vector time-series models. Similarly, it is difficult to know how to react to calibration and the display of favoured models, which again seem to have no probabilistic bases whatsoever. Calibration connected to qualitative information (assembled for US business cycles, e.g. by the National Bureau of Economic Research) may give an interesting framework for some discussions of short-term economic fluctuations, but this is distant from econometrics and not readily integrated with formal econometric research, except as a small part of post-model evaluations at each stage of a sequential process.

Two other aspects of basic time-series characterizations need to be identified, one associated with non-linear chaotic models (Frank and Stengos, 1988) and the other with dynamic specification and long-term economic relationships (Hendry *et al.*, 1984). Formal non-linear models have been part of economics for over fifty years, but their impact has recently re-emerged with a new recognition of chaos and its restrictions on the predictability of economic futures. Erratic patterns exhibited by random sequences are indistinguishable from the patterns generated by deterministic chaotic formulae.

Much of economic theory is static and dynamic specifications for econometric equations typically involve supplementary elements that are not easily drawn from theoretical considerations. Thus, the history of empirical research is full of instances in which some arbitrary 'adaptive', 'expectational' or 'error correction' adjustments are imposed on variables loosely identified by economic theory. Such adjustments have been enlarged by the concept of co-integration since it was first tentatively introduced by Sargan (1964) as a constraint on the joint long-run behaviour of wages and prices, and later made explicit by Granger (1981b, 1986) and Engle and Granger (1987) in more formal representations. The statistical complements of this concept, which links parallel movements in time series despite short-term divergence, are comprehensively described by Banerjee *et al.* (1993) and summarized by Ericsson (1991). The long-term connection between economic variables can be viewed as a modest stochastic form of dynamic equilibrium, with their temporary divergences being controlled by adjustment processes. In practice, co-integration modelling relies heavily on basic time-series models with a small degree of non-stationarity. Critical assessments of this economic interpretation and co-integration tests are offered by Kelly (1985) and Swamy *et al.* (1989), who question whether co-integration is indeed 'a property of the real world', but there can be little doubt that the concept has swept to immense fashionability, at least for the present moment.

CONCLUDING REMARKS

Econometrics must contain major probabilistic elements and its future characteristics will surely depend on some well-established historical conventions and the success of recent attempts to amend these conventions by reference to the strong demands of sequential analysis, weak or whimsical bases of specification and statistical inference, criticism and robustness, and some tentative attempts to clarify dynamic adjustments with co-integration and other forms of non-stationarity. Past practice in econometric research has been dominated by procedures which are convenient and are made feasible by their unqualified acceptance and the availability of computing software for estimation, testing and simulation. The probabilistic elements in applied research are not accurately portrayed in most textbooks, which stress confirmatory motives rather than sensible exploratory ones. Thus, dialogue among econometricians is typically muddled and driven by an apparent excess of wishful thinking, with an incomplete and awkward treatment of most philosophical issues and little appreciation of important problems such as measurement error, systemic instability, non-linearities and the inappropriate references to probability concepts. Against this bleak picture, recent reappraisals indicate that sceptics have sought to reduce the persistent neglect of these problems and advocate alternative procedures which are less myopic and more directed at progressive research.

8

CONCLUDING REMARKS AND OVERVIEW

The reconciliation of familiar probabilistic notions with normal economic contexts and the visible actions of economic agents is often troublesome. Major difficulties stem from the historical specificity of economic events and from the structural instability produced when market conditions and individual circumstances are amended by unanticipated erratic shocks and frequent qualitative changes, as well as from the placing of probability calculations in recognizable mental processes, basic flows of information, and any dialogues occurring within the principal decision-making groups. Other major difficulties arise from an obvious lack of precision and the costs of both acquiring and processing suitable information to facilitate the use of probabilities – imprecision and costs which are not readily put into the abstract formulations of models with probabilistic elements, and which must effectively narrow the perceived range of applicability for such models to illuminate most economic phenomena. Probability has to be more carefully used or it has to be replaced by alternative notions (such as possibility, belief functions, fuzziness and potential surprise), which could offer a better fit to the economic context. We cannot presume that probability density functions, fixed parameters (e.g. means, variances and correlations) and simple stochastic processes will always provide a suitable means of expressing uncertain outcomes, incomplete information or ignorance. Instead, more scepticism needs to be expressed and alternative approaches vigorously sought and assessed for their potential usefulness.

The structure and content of our treatment of probability are relatively straightforward. After a brief introduction, the first three chapters deal with primary visions and fundamental concepts of probability, utility and rationality – usually acknowledging the normal conventions of recent economic thought and the practices of economists and other social scientists. Chapter 1 surveys the variety in our understanding of probability by reference to axioms and normative issues. Chapter 2 focuses attention on expected utility as part of descriptive and normative bases for abstract choice under uncertainty as embedded in treatments of fair gambles, moral expectation, risk aversion and game theories. Since reliance on expected utility is quite pervasive throughout much of current economic theory, the primary problem confronted in this chapter is the degree to which expected utility is (and should be) used to guide actual choice in uncertain situations. Chapter 3 discusses various notions of rationality in terms of equivalents, portfolios, expectations, increasing risk and decision-making. Here, rationality is examined in the light of behavioural definitions, the search for self interest through explicit optimization, simple decision rules, theoretical 'applications, and internal consistency within axiomatic models. Some of the major technical, psychological and behavioural difficulties which caused certainty equivalents to fall from favour are identified. This partial dismissal provides a good background for a greater awareness of cognition amd mental aspects of choices and decisions – an awareness that has generalized expected utility and clarified theoretical paradoxes.

The remaining four chapters consider both probability and its principal alternatives in relation to ignorance and vagueness, new approaches to experimentation, the radical reformulation of beliefs and support, and econometric methods. Our terse treatment of such topics seeks to introduce unconventional ideas which seem to have some potential to affect the fashionable ways in which economists view uncertainty, structural instabilities and imprecision. Chapter 4 examines the ignorance of economic agents as reflected through imagination, inexactness and fuzzy set concepts, the appraisal of possibilities in various ways, and relative attitudes to acceptable hazards. Controversy emerges from the perception that ignorance is the state of not having sufficient information to facilitate use of a stable probability density function. Chapter 5 looks at some recent findings from economic experiments, which have strongly stimulated and supported interest in experimental irrationality, heuristics and routine rules for conduct, contextual pressures for choices and dynamic factors. Theories of expected utility and the paradoxes associated with them have provoked the design of new laboratory-style experiments to clarify, confirm or confront past promotion of expected utility. Such experiments can permit the numerical calibration of parameters and the ready manipulation of control factors, and their findings are difficult to generalize due to (perhaps excessive) specificity and restrictive precision, but this experimentation now seems to have become a useful addition to the normal tool-kit of economists and psychologists interested in choice and decision-making.

Chapter 6 takes up visions of unreliable and indeterminate probabilities by reference to personal circumstances and changing knowledge, belief functions and dynamic epistemic states, vagueness and incompleteness – before concluding with a brief recognition of the practical awkwardness for elucidation of personal probabilities and their practical exchange (and convergent modification) within decision-making groups through Bayesian dialogue, normally a social process rather than just a technical one which can be summarized in a simple mathematical format. Chapter 7 is devoted to the basic conventions of econometrics and their radical reappraisal over the last two decades, when the familiar structural or 'probabilistic' approach to estimation and inference (launched by some prominent scholars at the Cowles Commission in the 1940s and 1950s) came to be actively questioned. Our account seeks to clarify the unsettled state of econometrics, including the haphazard use of probability and a strong discontent, which is reflected in the advocacy of new cumulative research strategies to encourage more 'progress' or the meaningful combination of diverse empirical findings.

This terse but wide-ranging account of the uses of probability in relation to economic theory and the fashionable presumptions of economists' behaviour is driven by our firm view that most methods involving probability and the conclusions emerging from their use *always* have to be appraised, because suitable elements of economic theories and inferential procedures will frequently vary as actual and hypothetical circumstances change. Presumption of the evident appropriateness of one methodological approach, namely that which involves a fully specified and stable probability density function, is not sensitive to the economic context and needs to be augmented or replaced by a much greater reliance on other pragmatic methods. We accept the proposition that any real decision-makers will often settle for less precise answers and more plausible or 'realistic' explanations for the current and future characteristics of economic phenomena. This acceptance is not weakened by an acute awareness that many government agencies, professional organizations, public-interest groups and planning departments in large corporations use (and will continue to use) the presumption of probability density functions and

corresponding calculations for estimated parameters and corresponding predictions for both their prospective planning and retrospective
valuation. Our discussion identifies major areas where probability has
been a primary constituent of economic theory – e.g. hypothetical
betting, risk assessment, games, asset portfolios and data analysis in
econometrics – as well as wider illustrations such as environmental
epidemiology.

This book was designed to evaluate the feasibility of applying
common probabilistic techniques and to appraise the value of using
such techniques. Our evaluation is affected by a desire to evoke the
evident complexities of informational sources in determining numerical probabilities, mediated through individual efforts and social or
collective networks. Probabilistic techniques must be appreciated as
metaphors for thinking within individual minds and for group
reactions. Thus, the sensible discussions of fundamental issues (such
as the logical consistency of axioms for preference and coherence of
personal probabilities) should recognize the need to justify probability
as metaphor. Also, serious reconsideration of multi-stage models of
search, behavioural heuristics and sequential models of decision trees
seems necessary so as to determine their relative merit for the
economic context and any inter-connectedness with other conceptions
of human mental processes, which might fit that context too.

Another aspect of evaluation is the recognition of alternative and
more tentative concepts, such as fuzzy possibilities, beyond the conventional probability ones. The lure of modest proposals for a mathematization of intuitive possibilities and demonstrations of their
relative effectiveness in modelling uncertain conditions in particular
circumstances where the quantitative and qualitative forms of probability have been less successful, ought not to be resisted. Potentially
useful elements in determining numerical probabilities and choosing
among these and unconventional concepts must include expertise (the
ability to address changing situations by producing imaginative
options) and experience (the ability to integrate reactions to past
referential events within the current decision-making patterns), as
well as naturalistic approaches which allow for learning as choices
emerge, situations are clarified, and decisions are deferred, modified
and distributed over time.

A major conclusion of our efforts is the need for economists and other
social scientists to acquire a pluralistic approach to uncertainty and
probability. In expressing this conclusion, we have in mind the
misconceptions of both trained experts and naive subjects alike. The

optimal modelling procedure can involve many ways of determining probabilities, imagination, fuzziness, surprise and heuristics – or none of these alternatives but rather other approaches yet to be identified. We must appreciate that certain models which are sound for simple and well-defined problems may be quite misleading and costly when applied to more complex and ill-defined problems, so a certain degree of restraint is necessary in the promotion of any specific probability models for a wide variety of situations.

Some prominent views of probability can be crudely separated into two broad realistic and theoretical categories, depending on the relative amount of attention given to a perceived congruence with the observable features of actual human behaviour. Basically, realistic views judge quality by conformity with behaviour, while the proponents of theoretical approaches view any non-conformity as an indication of the need for individuals to change their behaviour so that it corresponds more closely to theoretical presumptions for rationality. We identify Keynes, Hicks, Shackle, Hayek, Simon and Tversky as illustrative of the rich variety found in the first of these categories. The contributions of Savage, Morgenstern, Machina and Lindley to economics and statistics represent views found in the second normative category.

Keynes's constructive theory involves logical propositions in the form of rational beliefs. The propositions are derived from knowledge and experience, and beliefs or objective measures exist by virtue of observed behaviour. His perspective entails major restrictions due to normal limitations on the acquisition of basic information as affected by misperceptions, incomparabilities and frequent non-quantification. For Keynes, sufficient congruence with observable human behaviour was an initial criterion without which meaningful probabilistic outcomes could not result. Hicks's perspective was dominated by reservations about the connection of frequentist notions with the 'state of the world', so he had little faith in the ability of most probability models to offer realistic information. In portfolio analysis and his treatment of money, he discounted the possibility of repetition and noted that situations would occur for which the available evidence may be insufficient to support probabilities. Ergodic or measure-preserving assumptions, he observed, are not compatible with historical conditioning and topicality so particular episodes, experiences and shocks cannot be addressed by familiar models containing the assumptions. From early acceptance of the two-parameter characterizations of uncertainty, Hicks moved to an obstinate and persistent concern with the need to modify models, adding parameters, rejecting ubiquitous

161

normality, and introducing more awareness of myopic information while keeping a basic historical perspective.

Shackle insisted on an even more radical dependence on unique cases and realistic elements for describing decisions, decision-making, the listing of limited options and potential outcomes of feasible actions. Thus, sensible models, from his kaleidic vision, must accommodate the explicit recognition of mental activities, creative imagination and the pressures of 'unknowledge' with many decisions reflecting uniqueness, immediacy, unpredictability and self-destructive aspects – clearly a firm rejection of frequentist probabilities, abstractions of static equilibrium theory and the use of mathematical expectations in favour of newer conceptions of potential surprise, imagined sequels, myopic focus and ascendancy, epistemic intervals and disbelief.

Against these perspectives driven by realistic elements, the promotion of a few normative axioms for rational choice by von Neumann and Morgenstern in the context of economic games (and more widely by Savage) raises the spectre of excessive precision for numerical probabilities, revives cardinal utility in a new form, expresses irrationality as violations of presumed transitivity and other constraints on individual preferences and the comparisons of well-established options, and offers much clearer foundations for any subjective or personal probabilities with a neo-Bernoullian basis. The strong attachment to economic games, expected utility and optimization is very attractive to economic theorists despite the awkward lack of descriptive power and the atemporal nature of most rational assumptions.

Savage's development of subjective expected utility theory involves attaching a strong variety of normative rationality to ideally coherent persons, with coherence defined in terms of the avoidance of a Dutchbook situation and the fulfilment of a given collection of probability axioms. He was relatively unfazed by the clear inadequacies of theoretical prescriptions to describe actual preferences, dismissing major issues of aggregation and consensus formation within groups by pointing to the potential occurrence of common values and similarities in preferences. However, he was very successful in widening the search for axiomatic characterizations of incomplete information and rational beliefs, and the Bayesian approach to statistical inference which he promoted continues to attract more followers. Higher-order probabilities, recognized by other prominent statisticians such as Good, were avoided by Savage as impractical – in favour of permitting some probabilities to be merely unsure. Among economists, the surveys of neo-Bernoullian theories by Machina illustrate the persisting attractions of firm probabilities bound to expected utility and its generalizations, with any criticisms of the independence axiom (and other axioms) and an interest in non-linear preferences being expressed through an active search for some alternative axioms rather than by the dismissal of the theoretical EU-framework itself.

Our book offers a modest introduction to the many approaches within which probability is integrated in economic theory and to methodological issues found in fitting fashionable probabilistic notions to the economic context. Brief accounts are provided for the contributions of the Friedman–Savage papers, moral expectation and expected-utility theories, and new views involving fuzziness, possibilities, experimentation and a naturalistic approach. All of these accounts need to be supplemented by reading some of the many references that are cited in the text. Similarly, rationality in choice is tentatively connected to general equilibrium models with stochastic elements, experimental validation and learning, but we fail to deal adequately with approaches to equilibrium solutions for noncooperative games. Other omissions include search theory, market processes, national accounts and sampling, the development of new dynamic-utility theories (by Edwards, Luce, Herrnstein and others), causality and statistics.

In our introduction, we promised to indicate the references that we favour to describe probability and the alternative concepts noted in other chapters. Inevitably our choice reflects personal interests in the history of probability, econometrics and economic theory. Our favourite books on probability include Daston (1988), Fine (1973), Hacking (1965, 1975), Keynes (1921), Maistrov (1974), Walley (1991), Kruger et al. (1987a, b), as well as the collections of papers offered by Edwards (1992b), Gardenfors and Sahlin (1988), Kahneman et al. (1982), Good (1983), and Moser (1990). A second tier with close connections to economics and social issues includes Ozga (1965), Fellner (1965), Hacking (1990), von Winterfeldt and Edwards (1986), Allais and Hagen (1979) and other menmbers of the series on expected utility issued by Reidel. Beyond these lists, many articles and papers offer enjoyable reading and illumination – Bunge (1989, 1995) and anything written by Simon, Shackle and Lindley, as well as relevant contributions to the *New Palgrave*. There is an exhausting supply of literature to stimulate the appetite for a better understanding of real economic phenomena and the ways in which economists and their scientific colleagues have sought to represent actual and 'rational' behaviour.

163

BIBLIOGRAPHY

Ajzen, I. (1977) 'Intuitive theories of events and the effects of base-rate information on prediction', *Journal of Personality and Social Psychology*, 35: 303–14.

Alchourron, C., Gardenfors, P. and Makinson, D. (1985) 'On the logic of theory change', *Journal of Symbolic Logic*, 50: 510–30.

Aldrich, J. (1989) 'Autonomy', *Oxford Economic Papers*, 41: 15–34.

Allais, M. (1952) 'The foundations of a positive theory of choice involving risk and a criticism of the postulates and axioms of the American school', English translation appearing in M. Allais and O. Hagen (eds) (1979) *Expected Utility Hypotheses and the Allais Paradox* (Boston: Reidel) 27–145.

—— (1984) 'The foundations of the theory of utility and risk', in O. Hagen and F. Wenstop (eds) *Progress in Utility and Risk Theory* (Dordrecht: Reidel) Part I.

—— (1990) 'Allais paradox', in J. Eatwell *et al.* (eds) *Utility and Probability* (London: Macmillan) 3–9.

Allais, M., and Hagen, O. (eds) (1979) *Expected Utility Hypotheses and the Allais Paradox* (Dordrecht: Reidel). Reviewed by G.R. Shafer (1984) *Journal of the American Statistical Association*, March, 224–35.

Anderson, R.G. and Dewalt, W.G. (1994) 'Replication and scientific standards in applied economics a decade after the Journal of Money, Credit and Banking Project', *Review, Federal Reserve Bank of St. Louis*, 76 (6): 79–83.

Ando, A., Fisher, F.M. and Simon, H.A. (1963) *Essays on the Structure of Social Science Models* (Cambridge, MA: MIT Press).

Appleby, L., and Starmer, C. (1987) 'Individual choice under uncertainty: a review of experimental evidence, past and present', in J.D. Hey and P.J. Lambert (eds) *Surveys in the Economics of Uncertainty* (Oxford: Basil Blackwell) ch. 2.

Arkes, H.R., and Hammond, K.R. (1986) *Judgment and Decision Making* (Cambridge: Cambridge University Press).

Arrow, K.J. (1951) 'Alternative approaches to the theory of choice in risk-taking situations', *Econometrica*,

—— (1965) 'The theory of risk aversion', in K.J Arrow (ed.) *Aspects of the Theory of Risk-Bearing* (Oxford: Basil Blackwell) Lecture Two.

164

Atkinson, A.C. (1986) *Plots, Transformations and Regression* (Oxford: Oxford University Press).

Bacharach, M. (1979) 'Normal Bayesian dialogues', *Journal of the American Statistical Association*, 74: 837–46.

Bancroft, T.A. (1944) 'On biases in estimation due to the use of preliminary tests of significance', *Annals of Mathematical Statistics*, 15: 190–204.

—— (1964) 'Analysis and inference for incompletely specified models involving the use of preliminary tests of significance', *Biometrics*, 20: 427–42.

Banerjee, A., Dolado, J., Galbraith, J.W. and Hendry, D.F. (1993) *Co-integration, Error Correction, and the Econometric Analysis of Non-stationary Data* (Oxford: Oxford University Press).

Bar-Hillel, M. (1980a) 'What features make samples seem representative?', *Journal of Experimental Psychology: Human Perception and Performance*, 6: 579–89.

—— (1980b) 'The base-rate fallacy in probability judgments', *Acta Psychologica*, 44: 211–33.

—— (1982) 'Studies of representativeness', in D. Kahneman, P. Slovic and A. Tversky (eds) *Judgment Under Uncertainty: Heuristics and Biases* (Cambridge: Cambridge University Press) ch. 5.

—— (1990) 'Back to base rates', in R.M. Hogarth (ed.) *Insights in Decision-Making* (Chicago: University of Chicago Press).

Baron, J. (1993) *Morality and Rational Choice* (Boston: Kluwer Academic Press).

Bartlett, M. (1975) *Probability, Statistics and Time* (London: Chapman and Hall).

Bassett, G.W. (1987) 'The St. Petersburg paradox and bounded utility', *History of Political Economy*, 19 (4): 517–23.

Bateman, B.W. (1987) 'Keynes's changing conception of probability', *Economics and Philosophy*, 3: 97–119.

—— (1990) 'The elusive logical relation: an essay on change and continuity in Keynes's thought', in D.E. Moggridge (ed.) *Perspectives on the History of Economic Thought. vol. IV: Keynes, Macroeconomics and Method* (Aldershot: Edward Elgar) ch. 5.

Bayarri, M.J. and DeGroot, M.H. (1989) 'Optimal reporting of predictions', *Journal of the American Statistical Association*, 84: 214–22.

Bell, D.E. (1982) 'Regret in decision making under uncertainty', *Operations Research*, 30: 961–81.

—— (1985) 'Disappointment in decision making under uncertainty', *Operations Research*, 33: 1–27.

Belsley, D.A., Kuh, E. and Welsch, R.E. (1980) *Regression Diagnostics* (New York: Wiley).

Bentzel, R. and Wold, H.O.A. (1946) 'On statistical demand analysis from the viewpoint of simultaneous equations', *Skandinavisk Aktuarietidskrift*, 95–114.

Berger, J.O. (1986) 'Comment', *Statistical Science*, 1 (3): 351–2.

Bergin, J. (1989) 'We eventually agree', *Mathematical Social Sciences*, 17: 57–66.

165

Bernasconi, M. and Loomes, G. (1992) 'Failures of the reduction principle in an Ellsberg-type problem', *Theory and Decision*, 32: 77–100.

Bernoulli, D. (1954) 'Exposition of a new theory on the measurement of risk' (English translation of 1738 paper), *Econometrica*, 22: 23–6.

Bezdek, J. (1993) 'Editorial. Fuzzy models – what are they and why?', *IEEE Transactions on Fuzzy Systems*, 1 (1): 1–5.

Bhaskar, R. (1989) *Reclaiming Reality* (London: Verso).

Bigg, R.J. (1990) *Cambridge and the Monetary Theory of Production* (New York: St Martin's Press), ch. 6.

Binmore, K. (1987) 'Experimental economics', *European Economic Review*, 257–64.

Bird, R.C., *et al.* (1965) '"Kuznets cycles" in growth rates: the meaning', *International Economic Review*, 6 (2): 229–39.

Black, R.D.C. (1990) 'Utility', in J. Eatwell, M. Milgate and P. Newman (eds) *Utility and Probability* (London: Macmillan) 295–302.

Blackwell, D. and Girshick, M.A. (1954) *Theory of Games and Statistical Decisions* (New York: Wiley).

Blume, L.E. and Easley, D. (1982) 'Learning to be rational', *Journal of Economic Theory*, 26: 340–51.

Borch, K.H. (1968) *The Economics of Uncertainty* (Princeton, NJ: Princeton University Press).

Bostic, R., Herrnstein, R.J. and Luce, R.D. (1990) 'The effect on the preference-reversal phenomenon of using choice indifferences', *Journal of Economic Behavior and Organization*, 13: 193–212.

Box, G.E.P. (1979) 'Some problems of statistics and everyday life', *Journal of the American Statistical Association*, 74: 1–4.

—— (1980) 'Sampling and Bayes' inference in scientific modelling and robustness', *Journal of the Royal Statistical Society*, A143, Part 4: 383–430.

Box, G.E.P. and Jenkins, G.M. (1976) *Time Series Analysis: Forecasting and Control* (San Francisco: Holden-Day).

Bray, M.M. and Savin, N.E. (1986) 'Rational expectations equilibria, learning and model specification', *Econometrica*, 54 (5): 1129–60.

Bunge, M. (1988) 'Two faces and three masks of probability', in E. Agazzi (ed.) *Probability in the Sciences* (Boston: Kluwer Academic Press) 27–49.

—— (1989) 'Game theory is not a useful tool for the political scientist', *Epistemologia*, 12: 195–212.

—— (1991) *A Critique of Rational Choice Theory* (Montreal: Foundations and Philosophy of Science Unit, McGill University).

—— (1995) 'The poverty of rational choice theory', in I.C. Jarvie and N. Laor (eds) *Critical Rationalism, Metaphysics and Science* (Boston: Kluwer Academic Press) vol. 1, 149–68.

Butler, D. and G. Loomes (1988) 'Decision difficulty and imprecise preferences', *Acta Psychologica*, 68: 183–96.

Bynum, W.F., Browne, E.J. and Porter, R. (eds) (1981) *Dictionary of the History of Science* (London: Macmillan).

Camerer, C.F. (1987) 'Do biases in probability judgment matter in markets? Experimental evidence', *American Economic Review*, 77 (5): 981–97.

—— (1992) 'Recent tests of generalizations of expected utility theory', in

W. Edwards (ed.) *Utility Theory: Measurements and Applications* (Dordrecht: Kluwer Academic Press), ch. 9.

Campbell, D.T. and Stanley, J.C. (1963) 'Experimental and quasi-experimental designs for research on teaching', in N.L. Gage (ed.) *Handbook of Research on Teaching* (Chicago: Rand McNally).

Carlson, J.B. (1987) 'Learning, rationality and the stability of equilibrium and macroeconomics', *Economic Review, Federal Reserve Bank of Cleveland*, Q4: 2–12.

Carnap, R. (1950) *Logical Foundations of Probability* (London: Routledge & Kegan Paul).

Carnap, R. and Jeffrey, R.C. (eds) (1971) *Studies in Inductive Logic and Probability* (Berkeley and Los Angeles: University of California Press) vol. 1.

Carter, C.F. (1957 a, b) 'A Revised Theory of Expectations' and 'The Present State of the Theory of Decisions Under Uncertainty', in C.F. Carter et al. (eds) *Uncertainty and Business Decisions* (Liverpool: Liverpool University Press) chs. V and XV.

—— (1972) 'On degrees Shackle: or, the meaning of business decisions', in C.F. Carter and J.L. Ford (eds) *Uncertainty and Expectations in Economics* (Oxford: Basil Blackwell) 30–42.

Carter, C.F. and Ford, J.L. (1972) *Uncertainty and Expectations in Economics* (Oxford: Basil Blackwell).

Carter, C.F., Meredith, G.P. and Shackle, G.L.S. (1957) *Uncertainty and Business Decisions* (Liverpool: Liverpool University Press).

Chamberlin, E.H. (1948) 'An experimental imperfect market', *Journal of Political Economy*, 56 (2): 95–108.

Charemza, W.W. and Deadman, D.F. (1992) *New Directions in Econometric Practice* (Aldershot: Edward Elgar).

Churchland, P.M. and Hooker, C.A. (eds) (1985) *Images of Science* (Chicago: University of Chicago Press).

Clark, J.M. (1947) 'Mathematical economists and others: a plea for communicability', *Econometrica*, 15 (2): 75–8.

Cohen, L.J. (1989) *An Introduction to the Philosophy of Induction and Probability* (Oxford: Clarendon Press) ch. 2.

Cole, S., Cole, J.R. and Dietrich, L. (1978) 'Measuring the cognitive state of scientific disciplines', in Y. Elkana et al. (eds) *Toward a Metric of Science: the Advent of Science Indicators* (New York: Wiley) 209–51.

Conroy, R.K., Swamy, P.A.V.B., Yanagida, J.F. and von zur Muehlen, P. (1984) *Agricultural Economics Research*, 36 (3): 1–15.

Cook, T.D. and Campbell, D.T. (1976) 'The design and conduct of quasi-experiments and true experiments in the field settings', in M.D. Dunnette (ed.) *Handbook of Industrial and Organizational Psychology* (Chicago: Rand McNally).

—— (1979) *Quasi-Experimentation. Design and Analysis Issues for Field Settings* (Chicago: Rand McNally).

Cook, R.D. and Weisberg, S. (1982) *Residuals and Influence in Regression* (London: Chapman and Hall).

Cooper, G.F. (1989) 'Current research directions in the development of expert systems based on belief notions', *Applied Stochastic Models and Data Analysis*, 5: 39–52.

Cox, D.R. (1977) 'The role of significance tests', *Scandinavian Journal of Statistics*, 4 (2): 49–70.

—— (1986) 'Some general aspects of the theory of statistics', *International Statistical Review*, 54 (2): 117–26.

Cramer, H. (1946) *Mathematical Methods of Statistics* (Princeton, NJ: Princeton University Press).

—— (1981) 'Mathematical probability and statistical inference: some personal recollections from an important phase of scientific development', *International Statistical Review*, 49: 309–17.

Darnell, A.C. and Evans, J.L. (1990) *The Limits of Econometrics* (Aldershot: Edward Elgar).

Daston, L. (1988) *Classical Probability in the Enlightenment* (Princeton, NJ: Princeton University Press).

Davidson, D. and Suppes, P. (1956) 'A finitistic axiomatization of subjective probability and utility', *Econometrica*, 24: 264–75.

Davis, D.D. and Holt, C.A. (1993) *Experimental Economics* (Princeton, NJ: Princeton University Press).

Dawid, A.P. (1982) 'The well-calibrated Bayesian', *Journal of the American Statistical Association*, 77: 605–14.

de Finetti, B. (1937) English translation in H.E. Kyburg and H.E. Smokler (eds) (1964) *Studies in Subjective Probability* (New York: Wiley).

—— (1972) *Probability, Induction and Statistics* (New York: Wiley).

—— (1974a) 'Bayesianism: its unifying role for both the foundations and applications of statistics', *International Statistical Review*, 42 (2): 117–30.

—— (1974b) 'The true subjective probability problem', in C.-A. S. Stael von Holstein (ed.) *The Concept of Probability in Psychological Experiments* (Dordrecht: Reidel) 15–23.

Debreu, G. (1986) 'Theoretic models: mathematical form and economic content', *Econometrica*, 54 (6): 1259–70.

DeGroot, M.H. (1974) 'Reaching a consensus', *Journal of the American Statistical Association*, 69: 118–21.

Dempster, A.P. (1967) 'Upper and lower probabilities induced by a multi-valued mapping', *Annals of Mathematical Statistics*, 38: 325–39.

—— (1968) 'A generalization of Bayesian inference', *Journal of the Royal Statistical Society*, B30: 205–47.

—— (1985) 'Probability, evidence and judgment', in J.M. Bernardo, M.H. DeGroot, D.V. Lindley and A.F.M. Smith (eds) *Bayesian Statistics 2* (Amsterdam: North-Holland) 119–32.

—— (1988) 'Probability, evidence and judgment', in D.E. Bell, H. Raiffa and A. Tversky (eds) *Decision Making* (Cambridge: Cambridge University Press) ch. 13.

Dobb, E.L. (1939) 'The length of the cycles which result from the graduation of chance elements', *Annals of Mathematical Statistics*, 10: 254–64.

Doob, J.L. (1961) 'Appreciation of Khinchin', in J. Neyman (ed.) *Proceedings of the Fourth Berkeley Symposium on Mathematical Statistics and Probability* (Berkeley and Los Angeles: University of California Press) vol.II.

—— (1976) 'Foundations of probability theory and its influence on the theory of statistics', in D.B. Owen (ed.) *On the History of Statistics and Probability* (New York: Marcel Dekker) 197–204.

Dreze, J.H. (1974) 'Axiomatic theories of choice, cardinal utility and subjective probability: a review', in J.H. Dreze (ed.) *Allocation Under Uncertainty: Equilibrium and Optimality* (New York: Wiley) 3–23.

Dubois, D. (1988) 'Possibility theory: searching for normative foundations', in B.R. Munier (ed.) *Risk, Decision and Rationality* (Dordrecht: Kluwer Academic Press) 601–14.

Dubois, D. and Prade, H. (1988a) 'Modelling uncertainty and inductive inference: a survey of recent non-additive probability systems', *Acta Psychologica*, 68: 53–78.

—— (1988b) *Possibility Theory* (New York: Plenum Press).

—— (1993) 'A glance at non-standard models and logics of uncertainty and vagueness', in J.-P. Dubucs (ed.) *Philosophy of Probability* (Boston: Kluwer Academic Press) ch. 9.

Durbin, J. (1987) 'Statistics and statistical sciences', *Journal of the Royal Statistical Society*, A150: 177–91.

—— (1988) 'Is a philosophical analysis for statistics attainable?', *Journal of Econometrics* (Annals) 37 (1): 51–62.

Eatwell, J., Milgate, M. and Newman, P. (eds) (1990a) *Utility and Probability* (London: Macmillan).

——(eds) (1990b) *Econometrics* (New York: Norton).

Edwards, W. (1968) 'Conservatism in human information processing', in B. Kleinmuntz (ed.) *Formal Representation of Human Judgment* (New York: Wiley).

—— (1990) 'Unfinished tasks: a research agenda for behavioral decision theory', in R.M. Hogarth (ed.) *Insights in Decision Making* (Chicago: University of Chicago Press) ch. 3.

—— (1992a) 'Toward the demise of economic man and woman; bottom lines from Santa Cruz', in W. Edwards (ed.) *Utility Theories: Measurements and Applications* (Boston: Kluwer Academic Press) ch. 10.

—— (ed.) (1992b) *Utility Theories: Measurements ans Applications* (Boston: Kluwer Academic Press).

Egidi, M. and Marris, R. (eds) (1992) *Economics, Bounded Rationality and the Cognitive Revolution* (Aldershot: Edward Elgar).

Einhorn, H.J. and Hogarth, R.M. (1981) 'Behavioral decision theory: processes of judgment and choice', *Annual Review of Psychology*, 32: 53–88.

—— (1985) 'Ambiguity and uncertainty in probabilistic inference', *Psychological Review*, 92: 433–61.

—— (1986) 'Decision making under ambiguity', *Journal of Business*, 59 (4, Part 2): S279–S250.

Ellsberg, D. (1954) 'Classical and current notions of "measurable utility"', *Economic Journal*, 64: 528–56.

—— (1961) 'Risk, ambiguity and the savage axioms', *Quarterly Journal of Economics*, 75: 643–69.

Engle, R.F. and Granger, C.W.J. (1987) 'Co-integration and error correction: representation, estimation and testing', *Econometrica*, 55: 251–76.

Engle, R.F., Hendry, D.F. and Richard, J.-F. (1983) 'Exogeneity', *Econometrica*, 51: 277–304.

Eppel, T., et al. (1992) 'Old and new roles for expected and generalized utility theories', in W. Edwards (ed.) *Utility Theories: Measurements and Applications* (Boston: Kluwer Academic Press) ch. 11.

Ericsson, N.R. (1991) *Cointegration, Exogeneity and Policy Analysis: an Overview*, International Finance Discussion Paper 415, Board of Governors of the Federal Reserve System, Washington.

Ericsson, N.R. and Hendry, D.F (1989) *Encompassing and Rational Expectations: How Sequential Corroboration Can Imply Refutation*, International Finance Discussion Paper 354, Board of Governors of the Federal Reserve System, Washington.

Ericsson, N.R., Campos, J. and Tran, H.-A. (1991) *PC-Give and Hendry's Econometric Methodology*, International Finance Discussion Paper 406, Board of Governors of the Federal Reserve System, Washington.

Evans, J.St B.T. (1989) *Bias in Human Reasoning: Causes and Consequences* (London: Lawrence Erlbaum Associates).

Feige, E.L. and D.K. Pearce (1979) 'The casual causal relationship between money and income: some caveats for time series analysis', *Review of Economics and Statistics*, 61; 521–33.

Feldstein, M.S. (1969) 'Mean-variance analysis in the theory of liquidity preference and portfolio selection', *Review of Economic Studies*, 36: 5–12.

Feller, W. (1945) 'Notes on the law of large numbers and "fair" games', *Annals of Mathematical Statistics*, 16: 301–4.

—— (1968) *An Introduction to Probability Theory and its Applications* (New York: Wiley) vol. 1: 3rd edition.

—— (1971) *An Introduction to Probability Theory and its Applications* (New York: Wiley) vol. 2: 2nd Edition.

Fellner, W. (1965) *Probability and Profit* (Homewood, IL: Irwin).

Fine, T.L. (1973) *Theories of Probability. An Examination of Foundations* (New York: Academic Press).

Fischhoff, B. (1980) 'For those condemned to study the past: heuristics and biases in hindsight', in R.A. Shweder and D.W. Fiske (eds) *New Directions for Methodology of Behavioral Science: Fallible Judgment in Behavioral Research* (San Francisco: Jossey-Bass).

Fischhoff, B. and Bar-Hillel, M. (1984) 'Diagnosticity and the base-rate effect', *Memory and Cognition*, 12: 402–10.

Fischhoff, B. and Beyth, R. (1975) '"I knew it would happen": remembered probabilities of once-future things', *Organizational Behavior and Human Performance*, 13: 1–16.

Fischhoff, B., et al. (1982) *Acceptable Risk* (Cambridge: Cambridge University Press).

Fishburn, P.C. (1970) *Utility Theory for Decision Making* (New York: Wiley).

—— (1982) 'Nontransitive measurable utility', *Journal of Mathematical Psychology*, 26: 31–67.

—— (1986) 'The axioms of subjective probability', *Statistical Science*, 1: 335–58.

—— (1991) 'On the theory of ambiguity', *International Journal of Information and Management Sciences*, 2 (2): 1–16.

—— (1993) 'The axioms and algebra of ambiguity', *Theory and Decision*, 34: 119–37.

Fisher, F.M. (1989) 'Games economists play: a non-cooperative view', *The Rand Journal of Economics*, 20: 113– 24.

Fisher, I. (1918) 'Is "utility" the most suitable term for the concept it is used to denote?', *American Economic Review*, 8: 335–7.

—— (1927) 'A statistical method of measuring "marginal utility" and testing the justice of a progressive income tax', in J.II. Hollander (ed.) *Economic Essays Contributed in Honor of John Bates Clark* (New York: Macmillan) 157–93.

Ford, J.L. (1983) *Choice, Expectation and Uncertainty* (Oxford: Basil Blackwell).

—— (1987) *Economic Choice Under Uncertainty* (Aldershot: Edward Elgar).

—— (1989) 'G.L.S. Shackle, 1903–', in D. Greenaway and J.R. Presley (eds) *Pioneers of Modern Economics in Britain* (New York: St Martin's Press) vol. 2, ch. 2.

Fox, K.A. (1982) 'Structural analysis and the measurement of demand for farm products: foresight, insight and hindsight in the choice of estimation techniques', in R.H. Day (ed.) *Economic Analysis and Agricultural Policy* (Ames: The Iowa State University Press) ch. 10.

—— (1989) 'Agricultural economists in the econometric revolution: institutional background, literature and leading figures', *Oxford Economic Papers*, 41: 53–70.

Frank, M. and Stengos, T. (1988) 'Chaotic dynamics in economic time-series', *Journal of Economic Surveys*, 2; 103–33.

Freixas, X. (1990) 'Certainty equivalent', in J. Eatwell, M. Milgate and P. Newman (eds) *Utility and Probability* (London: Macmillan) 19–21.

French, S. (1985) 'Group consensus probability distributions: a critical survey', in J.M. Bernardo, M.H. DeGroot, D.V. Lindley and A.F.M. Smith (eds) *Bayesian Statistics 2* (Amsterdam: North-Holland) 183–201.

Friedman, D. and Sundar, S. (1994) *Experimental Methods* (Cambridge: Cambridge University Press).

Friedman, M. (1949) 'Discussion', *American Economic Review*, 39 (3): 196–9.

—— (1952) Contribution to (ed.) *Wesley Clair Mitchell, The Economic Scientist* (New York: National Bureau of Economic Research).

Friedman, M. and Savage, I.J. (1948) 'The utility analysis of choices involving risk', *Journal of Political Economy*, 56 (4): 279–304.

—— (1952) 'The expected-utility hypothesis and the measurability of utility', *Journal of Political Economy*, 60 (6): 463–74.

Frisch, R. (1932) *New Methods of Measuring Marginal Utility* (Tubingen: Verlag von J.C.B. Mohr).

—— (1933a) 'Editorial', *Econometrica*, 1: 1–4.

—— (1933b) 'Propagation problems and impulse problems in dynamic equations', in *Economic Essays in Honor of Gustav Cassel* (London: George Allen and Unwin) 171–205.

Frisch, D. and Baron, J. (1988) 'Ambiguity and rationality', *Journal of Behavioral Decision Making*, 1: 149–57.

Frydman, R. and Phelps, E.S. (eds) (1983) *Individual Forecasting and Aggregate Outcomes* (Cambridge: Cambridge University Press).

Gaines, B.R. (1976) 'Foundations of fuzzy reasoning', *International Journal of Man-Machine Studies*, 8: 623–68.

Gardenfors, P. (1988) *Knowledge in Flux* (Cambridge, MA: MIT Press).

Gardenfors, P. and Sahlin, N.-E. (1982) 'Unreliable probabilities, risk taking and decision making', *Synthese*, 53: 361–86.

—— (1983) 'Unreliable Probabilities, Risk Taking and Decision Making', *Synthese*, 53: 361–386.

—— (eds) (1988) *Decision, Probability and Utility* (Cambridge: Cambridge University Press).

Genest, C. and Zidek, J.V. (1986) 'Combining probability distributions: a critique and an annotated bibliography', *Statistical Science*, 1 (1): 114–48.

Gibbons, J.D. and Pratt, J.W. (1975) 'P-values: interpretation and methodology', *The American Statistician*, 29: 20–5.

Gigerenzer, G. (1987) 'Probabilistic thinking and the fight against subjectivity', in L. Kruger, G. Gigerenzer and M.S. Morgan (eds) *The Probabilistic Revolution. vol. 2: Ideas in the Sciences* (Cambridge, MA: MIT Press) ch. 1.

Giles, R. (1976) 'Lukasiewicz logic and fuzzy set theory', *International Journal of Man-Machine Studies*, 8: 313–27.

—— (1982) 'Foundations for a theory of possibility', in M.M. Gupta and E. Sanchez (eds) *Fuzzy Information and Decision Processes* (Amsterdam: North-Holland) 183–95.

Glymour, C. (1983) 'On testing and evidence', in J. Earman (ed.) *Testing Scientific Theories* (Minneapolis: University of Minnesota Press) 3–16.

—— (1989) 'Comments on Lane and Cooper', *Applied Stochastic Models and Data Analysis*, 5: 83–8.

Goguen, J.A. (1967) 'L-fuzzy sets', *Journal of Mathematical Analysis and Applications*, 18: 145–74.

—— (1969) 'The logic of inexact concepts', *Synthese*, 19: 325–73.

Goldstein, M. (1981) 'Revising previsions: a geometric interpretation', *Journal of the Royal Statistical Association*, B43 (2): 105–30.

—— (1983) 'The prevision of a prevision', *Journal of the American Statistical Association*, 78: 817–19.

—— (1986) 'Exchangeable belief structures', *Journal of the American Statistical Association*, 81: 971–6.

Good, I.J. (1950) *Probability and the Weighing of Evidence* (London: Griffin).

—— (1957) 'Mathematical tools', in C.F. Carter, G.P. Meredith and G.L.S. Shackle (eds) *Uncertainty and Business Decisions* (Liverpool: Liverpool University Press) ch. 3.

—— (1962a) 'Subjective probability as the measure of a non-measurable set', in E. Nagel, P. Suppes and A. Tarski (eds) *Logic, Methodology and Philosophy of Science* (Stanford, CA: Stanford University Press) 319–29.

—— (1962b) 'How rational should a manager be?', *Management Science*, 8: 383–93.

—— (1965) *The Estimation of Probabilities* (Cambridge, MA: MIT Press).

—— (1971) 'The probabilistic explication of information, evidence, surprise, causality, explanation and utility', in V.P. Godambe and D.A. Sprott (eds) *Foundations of Statistical Inference* (Toronto: Holt, Rinehart and Winston).

—— (1981) 'Some logic and history of hypothesis testing', in J.C. Pitt (ed.) *Philosophy in Economics* (Boston: Reidel) 149–74.

—— (1983) *Good Thinking. The Foundations of Probability and its Applications* (Minneapolis: University of Minneapolis Press).

—— (1985) 'Weight of evidence: a brief survey', in J.M. Bernardo, M.H. DeGroot, D.V. Lindley and A.F.M. Smith (eds) *Bayesian Statistics 2* (Amsterdam: Elsevier) 249–70.

—— (1988) 'The interface between statistics and philosophy of science', *Statistical Science*, 3 (4): 386–412.

—— (1990) 'Subjective probability', in J. Eatwell et al. (eds) *Utility and Probability* (London: Macmillan) 255–69.

—— (1992) 'The Bayes/non-Bayes compromise: a brief review', *Journal of the American Statistical Association*, 87: 597–606.

Granger, C.W.J. (1980) 'Testing for causality – a personal viewpoint', *Journal of Economic Dynamics and Control*, 2: 329–52.

—— (1981a) 'The comparison of time series and econometric forecasting strategies', in J. Kmenta and J.B. Ramsey (eds) *Large-Scale Macro-Econometric Models* (Amsterdam: North Holland) 123–8.

—— (1981b) 'Some properties of time series data and their use in econometric model specification', *Journal of Econometrics*, 16: 121–30.

—— (1982) 'Generating mechanisms, models and causality', in W. Hildenbrand (ed.) *Advances in Econometrics* (Cambridge: Cambridge University Press) ch. 8.

—— (1986) 'Developments in the study of cointegrated economic variables', *Oxford Bulletin of Economics and Statistics*, 48: 213–28.

—— (1988) 'Some recent developments in a concept of causality', *Journal of Econometrics*, 39: 199–211.

—— (ed.) (1990) *Modelling Economic Series: Readings in Econometric Methodology* (Oxford: Oxford University Press).

Green, J. (1977) 'The nonexistence of informational equilibria', *Review of Economic Studies*, 44: 451–63.

Grether, D.M. and Plott, C.R. (1979) 'Economic theory of choice and the preferences reversal phenomenon', *American Economic Review*, 69 (4): 623–38.

Guesnerie, R. (1993) 'Successes and failures in coordinating expectations', *European Economic Review*, 37: 243–68.

Guttman, L. (1977) 'What is not what in statistics', *The Statistician*, 26 (2): 81–107.

—— (1985) 'The illogic of statistical inference for cumulative science', *Applied Stochastic Models and Data Analysis*, 1: 3–10.

Haavelmo, T. (1940) 'The inadequacy of testing dynamic theory by comparing theoretical solutions and observed cycles', *Econometrica*, 8: 312–21.

—— (1943a) 'Statistical testing of business-cycle theories', *Review of Economic Statistics*, 25: 13–18.

—— (1943b) 'The statistical implications of a system of simultaneous equations', *Econometrica*, 11: 1–12.

—— (1944) 'The probability approach to econometrics', Supplement to *Econometrica*, 12.

—— (1958) 'The role of the econometrician in the advancement of economic theory', *Econometrica*, 26: 351–7.

173

Hacking, I. (1965) *Logic of Statistical Inference* (Cambridge: Cambridge University Press).

—— (1975) *The Emergence of Probability* (Cambridge: Cambridge University Press).

—— (1990) *The Taming of Chance* (Cambridge: Cambridge University Press).

Hadar, J. and Russell, W. (1969) 'Rules for ordering uncertain prospects', *American Economic Review*, 59: 25–34.

Hagen, O. and Wenstop, F. (eds) (1984) *Progress in Utility and Risk Theory* (Boston: Reidel).

Hall, P. and Selinger, B. (1986) 'Statistical significance: balancing evidence against doubt', *Australian Journal of Statistics*, 28: 354–70.

Haltiwanger, J.C. and Waldman, M. (1989) 'Rational expectations in the aggregate', *Economic Inquiry*, 27: 619–36.

Hamaker, H.C. (1977) 'Bayesianism: a threat to the statistical profession?', *International Statistical Review*, 45: 111–15.

Hammond, K.R., et al. (1975) 'Social judgment theory', in M.F. Kaplan and S. Schwartz (eds) *Human Judgement and Decision Processes* (New York: Academic Press) 271–312.

Hammond, P. (1988a) 'Consequentialism and the independence axiom', in B.R. Munier (ed.) *Risk, Decision and Rationality* (Dordrecht: Reidel) 503–16.

—— (1988b) 'Consequentialist foundations for expected utility' *Theory and Decision*, 25: 25–78.

Hanoch, G. and Levy, H. (1969) 'The efficiency analysis of choices involving risk', *Review of Economic Studies*, 36: 335–46.

Harsanyi, J.C. (1977) *Rational Behavior and Bargaining Equilibrium in Games and Social Situations* (Cambridge: Cambridge University Press).

—— (1978) 'Bayesian decision theory and utilitarian ethics', *American Economic Review*, 68 (2): 223–8.

Hart, A.G. (1941) *Anticipations, Uncertainty and Dynamic Planning, Studies in Business Administration* (Chicago: University of Chicago Press) vol. 11.

—— (1942) 'Risk, uncertainty and the unprofitability of compounding probabilities', in O. Lange, F. McIntyre and T.O. Yntema (eds) *Studies in Mathematical Economics and Econometrics* (Chicago: University of Chicago Press) 110–18.

—— (1947) 'Keynes' Analysis of Expectations and Uncertainty', in S.E. Harris (ed.), *The New Economics: Keynes' Influence on Theory and Public Policy* (New York: Knopf) ch. 31.

Hartigan (1983) *Bayes Theory* (New York: Springer-Verlag) ch. 1.

Hayek, F.A (1960) *Constitution of Liberty* (London: Routledge & Kegan Paul).

—— (1964) 'The theory of complex phenomena', in M. Bunge (ed.) *The Critical Approach to Science and Philosophy. Essays in Honor of K.R. Popper* (New York: The Free Press).

—— (1978) 'Competition as a discovery procedure', *New Studies in Philosophy, Politics, Economics and the History of Ideas* (Chicago: University of Chicago Press) ch. 12.

—— (1986) 'The moral imperative of the market', in R. Harris *et al.* (eds)

The Unfinished Agenda (London: The Institute for Economic Affairs) 143–9.

Hendry, D.F. (1974) 'Stochastic specification in an aggregate demand model of the United Kingdom', *Econometrica*, 42 (3): 559–78.

—— (1980) 'Econometrics – alchemy or science?', *Economica*, 47: 387–406.

—— (1983a) 'On Keynesian model building and the rational expectations critique: a question of methodology', *Cambridge Journal of Economics*, 7: 69–75.

—— (1983b) 'Econometric modelling: the "consumption function" in retrospect', *Scottish Journal of Political Economy*, 30: 193–220.

—— (1987) 'Econometric methodology: a personal perspective', in T.F. Bewley (ed.) *Advances in Econometrics* (Cambridge: Cambridge University Press) vol. 2, ch. 10.

—— (1988) 'Encompassing', *National Institute Economic Review*, 125: 88–92.

—— (1992) 'Assessing empirical evidence in macroeconometrics with an application to consumers' expenditure in France', in A. Vercelli and N. Dimitri (eds) *Macroeconomics* (Oxford: Oxford University Press) ch. 13.

Hendry, D.F., Leamer, E.E. and Poirier, D.J. (1990) 'A conversation on econometric methodology', *Econometric Theory*, 6: 171–261.

Hendry, D.F., Pagan, A.R. and Sargan, J.D. (1984) 'Dynamic specification', in Z. Griliches and M.D. Intrigator (eds) *Handbook of Econometrics* (Amsterdam: North Holland) vol. 2, ch. 18.

Herstein, I.N. and Milnor, J. (1953) 'An axiomatic approach to measurable utility', *Econometrica*, 21: 291–7.

Hey, J.D. (1991) *Experiments in Economics* (Cambridge: Cambridge University Press).

Hicks, J.R. (1931) 'The theory of uncertainty and profit', *Economica*, 11: 170–89.

—— (1934) 'Application of mathematical methods to the theory of risk' (abstract) *Econometrica*, 2: 194–5.

—— (1935) 'A suggestion for simplifying the theory of money', *Economica*,

—— (1962) 'Liquidity', *Economic Journal*, 72: 787–802.

—— (1967) *Critical Essays in Monetary Theory* (Oxford: Clarendon Press) chs. 6 and 9.

—— (1977) *Perspectives*, ch. 8.

—— (1979) *Causality in Economics* (Oxford: Basil Blackwell).

—— (1982) *Money, Interest and Wages* (Oxford: Basil Blackwell).

Hicks, J.R. and Allen, R.G.D. (1934) 'A reconsideration of the theory of value', *Economica*, 1: 52–76, 196–219.

Hildenbrand, W. (1982) *Advances in Econometrics* (Cambridge: Cambridge University Press).

Hirshleifer, J. (1965) 'Investment decision under uncertainty – choice theoretic approaches', *Quarterly Journal of Economics*, 79 (4): 509–36.

—— (1966) 'Applications of the state preference approach', *Quarterly Journal of Economics*, 80 (2): 252–77.

Hodges, J.S. (1987) 'Uncertainty, policy analysis and statistics', *Statistical Science*, 2 (3): 259–91.

Hogarth, R.M. (1975) 'Cognitive processes and the assessment of subjective

175

probability distributions', *Journal of the American Statistical Association*, 70 (350): 271–94 (including comments by R.L. Winkler and W. Edwards).

—— (1987) *Judgement and Choice* (London: Wiley) 2nd edition.

Holgate, P. (1984) 'The influence of Huygens' work in dynamics on his contribution to probability', *International Statistical Review*, 52: 137–40.

Holt, C.A. (1986) 'Preference reversals and the independence axiom', *American Economic Review*, 76: 508–15.

Holt, C.C., Modigliani, F., Muth, J.F., and Simon, H.A. (1960) *Planning Production, Inventories and Work Force* (Englewood Cliffs, NJ: Prentice-Hall).

Hood, W.C. and Koopmans, T.C. (eds) (1953) *Studies in Econometric Method* (New York: Wiley).

Howard, R.A. (1992) 'In praise of the old time religion', in W. Edwards (ed.) *Utility Theories: Measurements and Applications* (Boston: Kluwer Academic Press) ch. 2.

Howrey, E.P, Klein, L.R. and McCarthy, M.D. (1974) 'Notes on testing the predictive performance of econometric models', *International Economic Review*, 15: 366–83.

Ingrao, B. and Israel, G. (1990) *The Invisible Hand* (Cambridge, MA: MIT Press).

Jacobs, R.L., Leamer, E.E. and Ward, M.P. (1979), 'Difficulties with testing for causation', *Economic Inquiry*, 17: 401–13.

Jeffreys, H. (1948) *Theory of Probability* (Oxford: Oxford University Press).

Jensen, N.E. (1967) 'An introduction to Bernoullian utility theory: I. Utility functions', *Swedish Journal of Economics*, 659: 163–83.

Joreskog, K.G. and Wold, H.O.A. (1982) *Systems Under Indirect Observation* (Amsterdam: North Holland).

Jorland, G. (1987) 'The Saint Petersburg paradox 1713–1937', in L. Kruger et al. (eds) *The Probabilistic Revolution* (Cambridge, MA: MIT Press) vol. 1, ch. 7.

Judge, G.C. and Bock, M.E. (1978) *The Statistical Implications of the Pre-Test and Stein-Rule Estimators in Econometrics* (Amsterdam: North Holland).

Judge, G.C., Griffiths, W.E., Hill, R.C. and Lee, T.-C. (1980) *The Theory and Practice of Econometrics* (New York: Wiley).

Jungermann, H. and de Zeeuw, G. (1977) *Decision-Making and Change in Human Affairs* (Dordrecht: Reidel) 275–324.

Kadane, J.B. (1992) 'Healthy scepticism as an expected-utility explanation of the phenomena of Allais and Ellsberg', *Theory and Decision*, 32: 57–64.

Kagel, J. and Roth, A.E. (eds) (1993) *Handbook of Experimental Economics* (Princeton, NJ: Princeton University Press).

Kahneman, D. (1991) 'Judgment and decision making: a personal view', *Psychological Science*, 2 (3): 142–5.

Kahneman, D. and Tversky, A. (1972) 'Subjective probability: a judgment of representativeness', in C.-A.S. Stael von Holstein (ed.) *The Concept of Probability in Psychological Experiments* (New York: Academic Press) 25–48. More detailed version appears in *Cognitive Psychology* (1972) 3: 430–54.

—— (1979) 'Prospect theory: an analysis of decision under risk', *Econometrica*, 47 (2): 263–91.

Kahneman, D., Slovic, P. and Tversky, A. (eds) (1982) *Judgment Under*

Uncertainty; Heuristics and Biases (Cambridge: Cambridge University Press). Reviewed by G.R. Shafer (1984) *Journal of the American Statistical Association*, March, 224.

Kalman, R.E. (1980) 'A system-theoretic critique of dynamic economic models', *International Journal of Policy Analysis and Information Systems*, 4: 3–22.

—— (1982a) 'Identification from real data', in H. Hazewinkel and A.H.G. Rinnooy Kan (eds) *Current Developments in the Interface: Economics, Econometrics, Mathematics* (Dordrecht: Reidel).

—— (1982b) 'Identification and problems of model selection in econometrics', in W. Hildenbrand (ed.) *Advances in Econometrics* (Cambridge: Cambridge University Press) 169–207.

—— (1983) 'Identifiability and modeling in econometrics', in P.R. Krishnaiah (ed.) *Developments in Statistics* (New York: Academic Press) vol. 4, ch. 2.

Kandel, A. (1986) *Fuzzy Mathematical Techniques With Applications* (Reading, MA: Addison-Wesley).

Karni, E. and Z. Safra (1987a) '"Preference reversals" and the observability of preferences by experimental methods', *Econometrica*, 55: 375–85.

—— (1987b) 'Preference reversals and the theory of choice under risk', *Econometrica*, 55: 675–85.

Kaser, G. (1989) *Acceptable Nuclear Risk: Some Examples From Europe*, Working Paper 409, European University Institute, Florence.

Kaysen, C. (1946–1947) 'A revolution in eonomic theory?', *Review of Economic Studies*, 14 (2): 1–15.

Kelly, C.M. (1985) 'A cautionary note on the interpretation of long-run equilibrium solutions in conventional macro models', *Economic Journal*, 95: 1078–86.

Keynes, J.M. (1921) *A Treatise on Probability* (London: Macmillan). Reissued with 'Editorial Foreword' by R.B. Braithwaite (1973) in *The Collected Writings of John Maynard Keynes. Vol. 8: A Treatise on Probability* (London: Macmillan).

—— (1936) *The General Theory of Employment, Interest and Money* (London: Macmillan).

Klein, L.R. (1960) 'Single equation vs. equation system methods of estimation in econometrics', *Econometrica*, 28: 866–71.

—— (1982) 'Economic theoretic restrictions in econometrics', in G.C. Chow and P. Corsi (eds) *Evaluating the Reliability of Macro-Economic Models* (New York: Wiley) ch. 4.

—— (1991) 'Econometric contributions of the Cowles commission, 1944–47: a retrospective view', *Banca Nazionale del Lavoro*, 177: 107–17.

Klein, G.A., J. Orasanu, R. Calderwood and C.E. Zsambok (1993) *Decision Making in Action: Models and Methods* (Norwood, NJ: Ablex).

Knight, F.H. (1921) *Risk, Uncertainty and Profit* (Chicago: University of Chicago Press).

Knobloch, E. (1987) 'Emile Borel as a probabilist', in L. Kruger, L.J. Daston and M. Heidelberger (eds) *The Probabilistic Revolution* (Cambridge, MA: MIT Press) vol. 1, ch. 9.

177

Kolmogorov, A.N. (1933) *Grundbegriffe der Wahrscheinlichkeitre-rechnung* (Berlin).
—— (1956) *Foundations of the Theory of Probability* (New York: Chelsea) 2nd English edition of Kolmogorov (1933).
Koopman, B.O. (1940a) 'The bases of probability', *Bulletin of the American Mathematical Society*, 46: 763–74.
—— (1940b) 'The axioms and algebra of intuitive probability', *Annals of Mathematics*, 41: 269–92.
Koopmans, T.C. (1941) 'The logic of econometric business-cycle research', *Journal of Political Economy*, 49: 157–81.
—— (1945) 'Statistical estimation of simultaneous economic relations', *Journal of the American Statistical Association*, 40: 448–66.
—— (1947) 'Measurement without theory', *Review of Economic Statistics*, 29: 161–72. Reprinted with other comments in R.A. Gordon and L.R. Klein (eds) (1965) *Readings in Business Cycles* (Homewood, IL: Irwin).
—— (1949) 'Identification problems in economic model construction', *Econometrica*, 17: 125–44.
—— (ed.) (1950a) *Statistical Inference in Dynamic Economic Models* (New York; Wiley).
—— (1950b) 'When is an equation system complete for statistical purposes?', in T.C. Koopmans (1950a) ch. 17.
Kruger, L., Daston, L.J. and Heidelberger, M. (eds) (1987a) *The Probabilistic Revolution. Vol.1: Ideas in History* (Cambridge, MA: MIT Press).
Kruger, L., Gigerenzer, G. and Morgan, M.S. (eds) (1987b) *The Probabilistic Revolution. Vol. 2: Ideas in the Sciences* (Cambridge, MA: MIT Press).
Lachman, L:M. (1976) 'From Mises to Shackle: an essay on Austrian economics and the Kaleidic Society', *Journal of Economic Literature*, 14 (1): 54–62.
—— (1990) 'G.L.S. Shackle's place in the history of subjectivist thought', in S.F. Frowen (ed.) *Unknowledge and Choice in Economics* (New York: St Martin's Press) 1–8.
Lane, D.A. (1989) 'Subjective probability and causality assessment', *Applied Stochastic Models and Data Analysis*, 5: 53–76.
Lange, O. (1944) *Price Flexibility and Employment* (Principia Press) ch. 6.
LaValle, I.H. (1992) 'Small worlds and sure things: consequentialism by the back door', in W. Edwards (ed.) *Utility Theories: Measurements and Applications* (Boston: Dordrecht) ch. 5.
Lawson, T. (1987a) 'The relative/absolute nature of knowledge and economic analysis', *Economic Journal*, 97: 951–70.
—— (1987b) 'Uncertainty and economic analysis', *Economic Journal*, 95: 909–27.
—— (1988) 'Probability and uncertainty in economic analysis', *Journal of Post Keynesian Economics*, 11: 38–65.
—— (1989a) 'Realism and instrumentalism in the development of econometrics', *Oxford Economic Papers*, 41: 236–58.
—— (1989b) 'Abstraction, tendencies and stylised facts: a realist approach to economic analysis', *Cambridge Journal of Economics*, 13: 59–78.
Leamer, E.E. (1978) *Specification Searches: Ad Hoc Inference With Nonexperimental Data* (New York: Wiley).

—— (1983a) 'Model choice and specification analysis', in Z. Griliches and M.D. Intriligator (eds) *Handbook of Econometrics* (Amsterdam: North Holland) vol. 1, ch. 5.

—— (1983b) 'Let's take the con out of econometrics', *American Economic Review*, 73: 31–43.

—— (1985a) 'Self-interpretation', *Economics and Philosophy*, 1: 295–302.

—— (1985b) 'Sensitivity analyses would help', *American Economic Review*, 75: 308–13.

—— (1987) 'Econometric metaphors', in T.F. Bewley (ed.) *Advances in Econometrics* (Cambridge: Cambridge University Press) vol. 2, ch. 9.

Lehmann, E.L. (1959) *Testing Statistical Hypotheses* (New York: Wiley).

Leontief, W. (1971) 'Theoretical assumptions and nonobserved facts', *American Economic Review*, 61: 1–7.

Leplin, J. (ed.) (1984) *Scientific Realism* (Berkeley and Los Angeles: University of California Press).

Levi, I. (1972) 'Potential Surprise in the context of inquiry', in C.F. Carter and J.L. Ford (eds) *Uncertainty and Expectations in Economics* (Oxford: Basil Blackwell) 213–36.

—— (1974) 'On indeterminate probabilities', *Journal of Philosophy*, 71: 391–418.

—— (1979) 'Inductive appraisal', in P.D. Asquith and H.E. Kyburg (eds) *Current Research in Philosophy of Science* (The Philosophy of Science Association) 339–51.

—— (1986a) 'The paradoxes of Allais and Ellsberg', *Economics and Philosophy*, 2: 23–53.

—— (1986b) *Hard Choices* (Cambridge: Cambridge University Press).

Lichtenstein, S. and Slovic, P. (1971) 'Reversals of preferences between bids and choices in gambling decisions', *Journal of Experimental Psychology*, 89: 46–55.

Lichtenstein, S., Fischhoff, B. and Phillips, L.D. (1977) 'Calibration of probabilities: the state of the art', in H. Jungermann and G. de Zeeuw (eds) *Decision Making and Change in Human Affairs* (Dordrecht: Reidel).

Lindley, D.V. (1965) *Introduction to Probability and Statistics From a Bayesian Viewpoint. Part 2: Inference* (Cambridge: Cambridge University Press).

—— (1971) *Making Decisions* (New York: Wiley).

—— (1975) 'The future of statistics – a Bayesian 21st century', *Advances in Applied Probability*, 7: 106–15.

—— (1982a) 'The Bayesian approach to statistics', in J.T. de Oliveira and B. Epstein (eds) *Some Recent Advances in Statistics* (New York: Academic Press) ch. 4.

—— (1982b) 'Scoring rules and the inevitability of probability', *International Statistical Review*, 50: 1–26.

—— (1987) 'The probability approach to the treatment of uncertainty in artificial intelligence and expert systems'. *Statistical Science*, 2 (1): 17–24.

Lindley, D.V., Tversky, A. and Brown, R.V. (1979) 'On the reconciliation of probability assessments', *Journal of the Royal Statistical Society*, 142 (2): 146–80.

Lindman, H.R. (1971) 'Inconsistent preferences among gamblers', *Journal of Experimental Psychology*, 89: 390–97.

Lipsey, R.G., P.N. Courant and D.D. Purvis (1994) *Microeconomics* (New York: HarperCollins) 8th Canadian edition.

Litterman, R.B. (1984) 'Forecasting and policy analysis with Bayesian vector autoregression models', *Federal Reserve Bank of Minneapolis Quarterly Review*, Fall, 30–41.

Liu, T.-C. (1955) 'A simple forecasting model for the U.S. economy', *IMF Staff Papers*, 4: 434–66.

—— (1960) 'Underidentification, structural estimation and forecasting', *Econometrica*, 28: 855–65.

Loeve, M. (1955) *Probability Theory* (Princeton, NJ: Princeton University Press).

Loomes, G. (1988a) 'Further evidence of the impact of regret and disappointment in choice under uncertainty', *Economica*, 55: 47–62.

—— (1988b) 'When actions speak louder than prospects', *American Economic Review*, 78: 463–70.

Loomes, G. and R. Sugden (1982) 'Regret theory: an alternative theory of rational choice under uncertainty', *Economic Journal*, 92: 805–24.

—— (1984) 'The importance of what might have been', in O. Hagen and F. Wenstop (eds) *Progress in Utility and Risk Theory* (Dordrecht: Reidel) 219–35.

—— (1986) 'Disappointment and dynamic consistency in choice under uncertainty', *Review of Economic Studies*, 53: 271–82.

—— (1987a) 'Some implications of a more general form of regret theory', *Journal of Economic Theory*, 41: 270–87.

—— (1987b) 'Testing for regret and disappointment choice under uncertainty', *Economic Journal*, 97: 118–29.

Lopes, L.L. (1994) 'Psychology and economics: perspectives on risk, cooperation and the marketplace', *Annual Review of Psychology*, 45: 197–227.

Lovell, M. and Selover, D. (1994) 'Econometric software accidents', *Economic Journal*, 713–25.

Lucas, R.E. (1972) 'Econometric testing of the natural rate hypothesis', in O. Eckstein (ed.) *The Econometrics of Price Determination* (Washington, DC: Board of the Federal Reserve System).

—— (1981) *Studies in Business-Cycle Theory* (Cambridge, MA: MIT Press).

—— (1986) 'Adaptive behavior and economic theory', *Journal of Business*, 59 (4, Part 2): S401-25.

—— (1987) *Models of business cycles* (Oxford: Basil Blackwell).

Lucas, R.E. and Sargent, T.J. (1979) 'After Keynesian macroeconomics', *Federal Reserve Bank of Minneapolis Quarterly Review*, 3 (2).

—— (1981) *Rational Expectations and Econometric Practice* (Minneapolis: University of Minnesota Press).

Luce, R.D. and Raiffa, H. (1957) *Games and Decisions* (New York: Wiley).

McCafferty, S. and Driskill, R. (1980) 'Problems of existence and uniqueness in nonlinear rational expectations models', *Econometrica*, 48 (5): 1313–17.

McCallum, B.T. (1993) *Macroeconomics After Two Decades of Rational Expectations*, Working Paper No. 4367, National Bureau of Economic Research.

McClennen, E.F. (1990) *Rationality and Dynamic Choice* (Cambridge: Cambridge University Press).

McCloskey, D. (1985a) *The Rhetoric of Economics* (Madison: Universty of Wisconsin Press) chs. 8 and 9.

—— (1985b) 'The loss function has been mislaid: the rhetoric of significance tests', *American Economic Review*, 75: 201–5.

MacCrimmon, K.R. and Larsson, S. (1979) 'Utility theory: axioms versus "paradoxes"', in M. Allais and O. Hagen (eds) *Expected Utility Hypotheses and the Allais Paradox* (Boston: Reidel) 333–409.

Machina, M.J. (1982) 'A stronger characterization of declining risk aversion', *Econometrica*, 50 (4): 1069–79.

—— (1983) 'Generalized expected utility analysis and the nature of observed violations of the independence axiom', in B.P. Stigum and F. Wenstrop (eds) *Foundations of Utility and Risk Theory With Applications* (Dordrecht: Reidel) 263–93.

—— (1987) 'Choice under uncertainty: problems solved and unsolved', *Economic Perspectives*, 1 (1): 121–54.

—— (1989) 'Dynamic consistency and non-expected utility models of choice under uncertainty', *Journal of Economic Literature*, 27: 1622–8.

Machina, M.J. and Neilson, W.S. (1987) 'The Ross characterization of risk aversion: strengthening and extension', *Econometrica*, 55 (5): 1139–49.

Machina, M.J. and Rothschild, M. (1990) 'Risk' in J. Eatwell, M. Milgate and P. Newman (eds) *Utility and Probability* (London: Macmillan) 227–39.

McKelvey, R.D. and Page, T. (1986) 'Common knowledge, consensus and aggregate information', *Econometrica*, 54: 109–27.

MacKinnon, J.G. (1983) 'Model specification tests against non-nested alternatives', *Econometric Reviews*, 2: 85–110.

Maistrov, L.E. (1974) *Probability Theory: A Historical Sketch* (New York: Academic Press).

Malinvaud, E. (1969) 'First order certainty equivalence', *Econometrica*, 37: 706–18.

—— (1991) 'Econometric methodology at the Cowles Commission: rise and maturity', in K.J. Arrow et al. (eds) *Cowles Fiftieth Anniverary* (New Haven: The Cowles Foundation for Research in Economics) 49–79.

Mann, H.B. and Wald, A. (1943) 'On the statistical treatment of linear stochastic difference equations', *Econometrica*, 11.

Marales, J.-A. (1971) *Bayesian Full Information Structural Analysis* (Berlin: Springer-Verlag).

March, J.G. (1978) 'Bounded rationality, ambiguity and the engineering of choice', *The Bell Journal of Economics*, 9 (2): 587–680.

Markowitz, H.M. (1952a) 'The utility of wealth', *Journal of Political Economy*, 60: 151–8.

—— (1952b) 'Portfolio selection', *Journal of Finance*, 7: 77–91.

—— (1959) *Portfolio Selection* (New Haven: Yale University Press).

Marschak, J. (1950) 'Rational behaviour, uncertain prospects and measurable utility', *Econometrica*, 18: 111–41.

—— (1953) 'Economic measurements for policy and prediction', in W.C. Hood and T.J. Koopmans (eds) *Studies in Econometric Method* (New York: Wiley) ch. 1.

—— (1954) 'Probability in the social sciences', in P.F. Lazarsfeld (ed.) *Mathematical Thinking in the Social Sciences* (New York: Free Press) 166–215.

Marshall, A. (1920) *Principles of Economics* (London: Macmillan) 8th edition, Book III.

Meier, P. (1986) 'Damned liars and expert witnesses', *Journal of the American Statistical Association*, 81: 269–76.

Menger, K. (1967) 'The role of uncertainty in economics' (English translation of 1934 paper) in M. Shubik (ed.) *Essays in Honor of Oskar Morgenstern* (Princeton, NJ: Princeton University Press) ch. 16.

Menges, G. (1973) 'Inference and decision', in G. Menges et al. (eds) *Inference and Decision* (Toronto: University Press of Canada) 1–16.

Mills, E.S. (1957) 'The theory of inventory decisions', *Econometrica*, 25: 222–38.

—— (1962) *Price, Output and Inventory Policy* (New York: Wiley).

Mizon, G.E. (1984) 'The encompassing approach to econometrics', in D.F. Hendry and K.F. Wallis (eds) *Econometrics and Quantitative Economics* (Oxford: Basil Blackwell).

Mizon, G.E. and Richard, J.-F. (1986) 'The encompassing principle and its application to testing non-nested hypotheses', *Econometrica*, 54: 657–78.

Moore, H.L. (1908) 'The statistical complement of pure economics', *Quarterly Journal of Economics*, 23: 1–33.

—— (1917) *Forecasting the Yield and Price of Corn* (New York: Macmillan).

Morgan, M.S. (1990a) 'Statistics without probability and Haavelmo's revolution in econometrics', in L. Kruger, G. Gigerenzer and M.S. Morgan (eds) *The Probabilistic Revolution* (Cambridge, MA: MIT Press) vol. 2, ch. 8.

—— (1990b) *The History of Econometric Ideas* (Cambridge: Cambridge University Press).

Morgenstern, O. (1979) 'Some reflections on utility' in M. Allais and O. Hagen (eds) *Expected Utility Hypotheses and the Allais Paradox* (Boston: Reidel) 175–83.

Moser, P.K. (1990) *Rationality in Action* (Cambridge: Cambridge University Press).

Muth, J.F. (1961) 'Rational expectations and the theory of price movements', *Econometrica*, 29: 315–35.

Nagel, E. (1939) 'Principles of the theory of probability', *International Encyclopedia of Unified Science*, I, No. 6 (Chicago: University of Chicago Press).

Neyman, J. (1952) *Lectures and Conferences on Mathematical Statistics and Probability* (Washington: Graduate School of US Department of Agriculture) 2nd Edition.

—— (1957) '"Inductive behavior" as a basic concept of philosophy of science', *Review of the International Institute of Statistics*, 25: 7–22.

—— (1977) 'Frequentist probability and frequentist statistics', *Synthese*, 36: 97–131.

Neyman, J. and Pearson, E.S. (1933) 'On the problem of the most efficient tests of statistical hypotheses', *Philosophical Transactions of the Royal Society of London*, A231: 289–337. Also reprinted in *Joint Statistical Papers of J.*

Neyman and E.S. Pearson (1933) (Berkeley, CA: University of California Press) 140–85.

O'Donnell, R.M. (1990) 'Continuity in Keynes's conception of probability', in D.E. Moggridge (ed.) *Perspectives in the History of Economic Thought. Vol. IV: Keynes, Macroeconomics and Method* (Aldershot: Edward Elgar) ch. 4.

Ordeshook, P.C. (1986) *Game Theory and Political Theory: An Introduction* (Cambridge: Cambridge University Press).

Oskamp, S. (1965) 'Overconfidence in case-study judgments', *Journal of Consulting Psychology*, 29: 261–5.

Overman, E.S. (ed.) (1988) *Methodology and Epistemology for Social Science. Selected Papers of Donald T. Campbell* (Chicago: University of Chicago Press).

Ozga, S.A. (1965) *Expectations in Economic Theory* (London: Weidenfeld and Nicolson).

Pagan, A. (1987) 'Three econometric methodologies: a critical appraisal', *Journal of Economic Surveys*, 1 (1): 3–24.

Paque, K.-H. (1990) 'Pattern predictions in economics: Hayek's methodology of the social sciences revisited', *History of Political Economy*, 22 (2): 281–94.

Plott, C.R. (1982) 'Industrial organization and experimental economics', *Journal of Economic Literature*, 1485–527.

—— (1986) 'Rational choice in experimental markets', *Journal of Business*, 59 (4, Part2): S301–27.

—— (1987) 'Dimensions of parallelism: some policy applications of experimental methods', in A.E. Roth (ed.) *Laboratory Experimentation in Economics* (Cambridge: Cambridge University Press) ch. 7.

Pope, R. (1986) 'Consistency and expected utility theory', in L. Daboni, A. Montesano and M. Lines (eds) *Recent Developments in the Foundations of Utility and Risk Theory* (Dordrecht: Reidel) 215–29.

Pratt, J.W. (1964) 'Risk aversion in the small and in the large', *Econometrica*, 32: 122–36.

—— (1986) 'Comment', *Statistical Science*, 1 (4): 498–9.

Pratt, J.W. and Schlaifer, R. (1988) 'On the interpretation and observation of laws', *Journal of Econometrics*, 39: 23–52.

Pratt, J.W., Raiffa, H. and Schlaifer, R. (1964) 'The foundations of decision under uncertainty: an elementary exposition', *Journal of the American Statistical Association*, 59: 353–75.

Putnam, H. (1987) *The Many Faces of Realism* (La Salle, IL: Open Court).

Raiffa, H. (1968) *Decision Analysis* (Reading, MA: Addison-Wesley).

Ralescu, A.L. and Ralescu, D.A. (1984) 'Probability and fuzziness', *Information Sciences*, 34: 85–92.

Ramsey, F.P. (1926) 'Truth and probability'. Reprinted in R.B. Braithwaite (ed.) *The Foundations of Mathematics and Other Logical Essays* (London: Routledge & Kegan Paul).

Renyi, A. (1970) *Foundations of Probability* (San Francisco, CA: Holden-Day).

Roberts, F.S. (1979) *Measurement Theory* (Reading, MA: Addison-Wesley).

Rosenkrantz, R.D. (1977) *Inference, Method and Decision* (Boston: Reidel) ch.10.

Ross, M. and Sicoly, F. (1979) 'Egocentric biases in availability', *Journal of Personality and Social Psychology*, 37: 322–46.

Ross, S. (1981) 'Some stronger measures of risk aversion in the small and the large with applications', *Econometrica*, 49: 621–38.

Roth, A.E. (1986) 'Laboratory experimentation in economics', *Economics and Philosophy*, 2: 245–73.

—— (1987) *Laboratory Experimentation in Economics* (Cambridge: Cambridge University Press).

Rothschild, M. and Stiglitz, J.E. (1970) 'Increasing risk. I. A definition', *Journal of Economic Theory*, 2: 225–43.

Rottenberg, T.L. (1971) 'The Bayesian approach and alternatives to econometrics', in M.D. Intriligator (ed.) *Frontiers of Quantitative Economics* (Amsterdam: North Holland) 194–210.

—— (1973) *Efficient Estimation With A Priori Information* (New Haven: Cowles Foundation for Research in Economics).

Russo, J.E. and Schoemaker, P.J.H. (1989) *Decision Traps* (New York: Simon & Schuster).

Samuelson, P.A. (1950) 'Probability and the attempts to measure utility', *Economic Review* (Tokyo). English version given in J.E. Stiglitz (ed.) (1966) *The Collected Scientific Papers of Paul A. Samuelson* (Cambridge, MA: MIT Press) vol. 1, ch. 12.

—— (1977) 'St Petersburg paradoxes: defanged, dissected and historically described', *Journal of Economic Literature*, 15 (1): 24–55.

Samuelson, P.A. and Nordhaus, W.D. (1985) *Principles of Economics* (New York: McGraw-Hill) 12th edition.

Samuelson, P.A., Koopmans, T.C. and Stone, J.R.N. (1954) 'Report of the evaluative committee for *Econometrica*', *Econometrica*, 22.

Samuelson, W. and Zeckhauser, R. (1988) 'Status quo bias in decision making', *Journal of Risk and Uncertainty*, 1: 7–59.

Sargan, J.D. (1964) 'Wages and prices in the United Kingdom: a study in econometric methodology', in P.E. Hart, G. Mills and J.K. Whitaker (eds) *Econometric Analysis for National Economic Planning* (London: Butterworth).

Sargent, T.J. (1979) 'Estimating vector autoregressions using methods not based on explicit economic theories', *Federal Reserve Bank of Minneapolis Quarterly Review*, 3 (3): 8–15.

—— (1983) 'An economist's foreword', in P. Whittle (ed.) *Prediction and Regulation by Linear Least-Square Methods* (Minneapolis: University of Minnesota Press).

Sauermann, H. and Selten, R. (1959) 'Ein Oligolpolexperiment', *Zeitschrift fur die Gesamte Staatswissenschaft*, 115: 427–71.

Savage, L.J. (1954) *The Foundations of Statistics* (New York: Wiley). 2nd revised edition reissued in 1972 (New York: Dover).

—— (1961) 'The foundations of statistics reconsidered', in J. Neyman (ed.) *Proceedings of the Fourth Berkeley Symposium on Mathematical Statistics and Probability* (Berkeley and Los Angeles: University of California Press) vol. 1: 575–86.

—— (1972) *The Foundations of Statistics* (New York: Dover) 2nd edition.

—— (1977) 'The shifting foundations of statistics', in R.G. Colodny (ed.) *Logic, Laws and Life* (Pittsburgh: University of Pittsburg Press) 3–18.

Schilpp, P.A. (ed.) (1963) *The Philosophy of Rudolf Carnap* (LaSalle, IL: Open Court).

Schkade, D.A. and Johnson, E.J. (1989) 'Cognitive processes in preference reversals', *Organizational Behavior and Human Decision Processes*, 44: 203–31.

Schlaifer, R. (1959) *Probability and Statistics for Business Decisions* (New York: McGraw-Hill).

Schmeider, D. (1989) 'Subjective probability and expected utility without additivity', *Econometrica*, 57 (3): 571–87.

Schoemaker, P.J.H. (1982) 'The expected utility model: its variants, purposes, evidence and limitations', *Journal of Economic Literature*, 20 (2): 529–63.

Schotter, A. (1992) 'Oskar Morgenstern's contribution to the development of the theory of games', in E.R. Weintraub (ed.) *Toward a History of Game Theory* (Durham: Duke University Press) 95–112.

Schultz, H. (1938) *The Theory and Measurement of Demand* (Chicago; University of Chicago Press).

Sebenius, J. and Geanakoplos, J. (1983) 'Don't bet on it: contingent agreements with asymmetric information', *Journal of the American Statistical Association*, 78: 424–6.

Segal, U. (1988) 'Does the preference reversal phenomenon necessarily contradict the independence axiom', *American Economic Review*, 78 (1): 233–6.

Sen, A. (1986) 'Rationality and uncertainty', in L. Daboni et al. (eds) *Recent Developments in the Foundations of Utility and Risk Theory* (Boston: Reidel) 3–25.

Shackle, G.L.S. (1949) *Expectation in Economics* (Cambridge: Cambridge University Press).

—— (1958) *Time in Economics* (Amsterdam: North-Holland).

—— (1979a) 'On Hicks's causality in economics: a review article', *Greek Economic Review*, 1 (2): 43–55.

—— (1979b) *Imagination and the Nature of Choice* (Edinburgh: Edinburgh University Press).

—— (1986) 'Decision', *Journal of Economic Studies*, 13 (5): 58–62.

Shafer, G. (1976a) *A Mathematical Theory of Evidence* (Princeton, NJ: Princeton Unversity Press).

—— (1976b) 'A theory of statistical evidence', in W.L. Harper and C.A. Hooker (eds) *Foundations of Probability Theory, Statistical Inference and Statistical Theories of Science* (Dordrecht: Reidel) vol. 2, 365–436.

—— (1978) 'Non-additive probabilities in the work of Bernoulli and Lambert', *Archive for History of Exact Sciences*, 19: 309–70.

—— (1981) 'Constructive probability', *Synthese*, 48: 1–60.

—— (1982a) 'Belief functions and parametric models', *Journal of the Royal Statistical Society*, B44: 322–52.

—— (1982b) 'Lindley's paradox', *Journal of the American Statistical Association*, 77: 325–51.

—— (1985) 'Conditional probability', *International Statistical Review*, 53: 261–77.

185

—— (1986) 'Savage revisited', *Statistical Science*, 1 (4): 463–501, including comments and rejoinder.

—— (1987) 'Probability judgment in artificial intelligence and expert systems', *Statistical Science*, 2 (1): 3–16.

—— (1988) 'Savage revisited', in D.E. Bell, H. Raiffa and A. Tversky (eds) *Decision Making* (Cambridge: Cambridge University Press) ch. 10.

Shafer, G. and A. Tversky (1988) 'Languages and designs for probability judgment', in D.E. Bell, H. Raiffa and A. Tversky (eds) *Decision Making* (Cambridge: Cambridge University Press) ch. 11.

Sherman, R. (1974) 'The psychological difference between ambiguity and risk', *Quarterly Journal of Economics*, 88: 166–9.

Simon, H.A. (1952) 'On the application of servomechanism theory in the study of production control', *Econometrica*, 20: 247–68.

—— (1953) 'Causal ordering and identifiability', in W.C. Hood and T.C. Koopmans (eds) *Studies in Econometric Method* (New York: Wiley) ch. 3.

—— (1956) 'Dynamic programming under uncertainty with a quadratic criterion function', *Econometrica*, 24: 74–81.

—— (1966) 'Theories of decision-making in economics and behavioural science', in *Surveys of Economic Theory. Vol III* (London: Macmillan). Prepared for Royal Economic Society and American Economic Association.

—— (1976) 'From substantive to procedural rationality', in S. Latsis (ed.) *Method and Appraisal in Economics* (Cambridge: Cambridge University Press) 129–48.

—— (1977) *Models of Discovery* (Dordrecht: Reidel).

—— (1982) *Models of Bounded Rationality* (Cambridge, MA: MIT Press).

—— (1983) *Reason in Human Affairs* (Stanford, CA: Stanford University Press).

—— (1986) 'Rationality in Psychology and Economics', *Journal of Business*, 59: S209–24.

—— (1990) 'Invariants of human behavior', *Annual Review of Psychology*, 41: 1–19.

Simon, H.A. and C.A. Kaplan (1989) 'Foundations of cognitive science', in M.I. Posner (ed.) *Foundations of Cognitive Science* (Cambridge, MA: MIT Press) ch. 1.

Sims, C.A. (1972a) 'Are there exogenous variables in short-run production relations?', *Annals of Economic and Social Measurement*, 1 (1): 17–36.

—— (1972b) 'Money, income and causality', *American Economic Review*, 62: 540–52.

—— (1986) 'Are forecasting models usable for policy analysis?', *Federal Reserve Bank of Minneapolis Quarterly Review*, Winter: 2–16.

Sinn, H.-W. (1983) *Economic Decisions Under Uncertainty* (Amsterdam: North-Holland) ch. 2.

Slovic, P. and Lichtenstein, S. (1983) 'Preference reversals: a broader perspective', *American Economic Review*, 73: 596–605.

Slutsky, E. (1927) 'The summation of random causes as the source of cyclical processes', in *Problems of Economic Conditions* (Moscow: The Conjuncture Institute). Revised English translation is given in *Econometrica* (1937) 5: 105–46.

—— (1937) 'The summation of random causes as the source of cyclic processes', *Econometrica*, 5: 105–46. English translation of paper published in 1927.

Smets, S. (1982a) 'Subjective probability and fuzzy measures', in M.M. Gupta and E. Sanchez (eds) *Fuzzy Information and Decision Processes* (Amsterdam: North-Holland) 87–91.

—— (1982b) 'Probability of a fuzzy event: an axiomatic approach', *Fuzzy Sets and Systems*, 7: 153–64.

Smith, A.F.M. (1984) 'Present position and potential developments: some personal views', *Journal of the Royal Statistical Society*, A147, Part 2: 245–59.

Smith, C.A.B. (1961) 'Consistency in statistical inference and decision', *Journal of the Royal Statistical Society*, B23: 1–37.

Smith, V.L. (1982) 'Microeconomic systems as an experimental science', *American Economic Review*, 72 (5); 923–55.

—— (1985) 'Experimental economics: reply', *American Economic Review*, March.

—— (1989) 'Theory, experiment and economics', *Journal of Economic Perspectives*, 3 (1): 151–69.

—— (1991) *Papers in Experimental Economics* (Cambridge: Cambridge University Press).

Spanos, A. (1989) 'On rereading Haavelmo: a retrospective view of econometric modeling', *Econometric Theory*, 5: 405–29.

Spiegelhalter, D.J. (1987) 'Probabilistic expert systems in medicine: practical issues in handling uncertainty', *Statistical Science*, 2 (1): 25–30.

Stephen, F.H. (1986) 'Decision making under uncertainty: in defence of Shackle', *Journal of Economic Studies*, 13 (5): 45–57.

Stigler, G.J. (1950) 'The development of utility theory', *Journal of Political Economy*, 58: 307–27, 373–96. Reprinted in G.J. Stigler (1965) *Essays in the History of Economics* (Chicago: University of Chicago Press).

Stigum, B.P. and Wenstop, F. (eds) (1983) *Foundations of Utility and Risk Theory With Applications* (Boston: Reidel).

Strotz, R.H. (1953) 'Cardinal utility', *American Economic Review*, 43: 384–97.

Strotz, R.H. and Wold, H.O.A. (1960) 'Recursive vs. Nonrecursive systems: an attempt at synthesis', *Econometrica*, 28: 417–27.

Sugden, R. (1985) 'New developments in the theory of choice under uncertainty', *Bulletin of Economic Research*, 38: 1–24.

—— (1986) 'Regret, recrimination and rationality', in L. Daboni, A. Montesano and M. Lines (eds) *Recent Developments in the Foundations of Utility and Risk Theory* (Boston: Reidel) 67–80.

Suppes, P. (1956) 'The role of subjective probability and utility in decision-making', in J. Neyman (ed.) *Proceedings of the Third Berkeley Symposium on Mathematical Statistics and Probability* (Berkeley and Los Angeles: University of California Press) vol. 5, 61–73.

—— (1974) 'The measurement of belief', *Journal of the Royal Statistical Society*, B36: 160–75.

—— (1976) 'Testing theories and the foundations of statistics', in W.L. Hooker and W.L. Harper (eds) *Foundations of Probability Theory, Statistical Inference and Statistical Theories of Science* (Boston: Reidel) vol. II: 437–52.

Sutton, J. (1987) 'Bargaining experiments', *European Economic Review*, 31: 272–84.

—— (1990) 'Explaining everything, explaining nothing? Game theoretic models in industrial organisation', *European Economic Review*, 34: 505–12.

—— (1993) 'Much ado about auctions', *European Economic Review*, 37: 317–19.

Swamy, P.A.V.B., Conway, R.K. and von zur Muehlen, P. (1985) 'The foundations of econometrics – are there any?', *Econometric Reviews*, 4 (1): 1–61.

Swamy, P.A.V.B., von zur Muehlen, P. and Mehra, J.S. (1989) *Co-integration: Is It a Property of the Real World?*, Finance and Economics Discussion Paper 96, Federal Reserve Board, Washington.

Taylor, J.B. (1977) 'Conditions for unique solutions in stochastic macro-economic models with rational expectations', *Econometrica*, 45: 1377–86.

—— (1985) 'Rational expectations models in macroeconomics', in K.J. Arrow and S. Honkakapoja (eds) *Frontiers of Economics* (Oxford: Basil Blackwell).

Taylor, S.E. (1982) 'The availability bias in social perception and interaction', in D. Kahneman, P. Slovic and A. Tversky (eds) *Judgment Under Uncertainty: Heuristics and Biases* (Cambridge: Cambridge University Press) ch. 13.

Terano, T., Asai, K. and Sugeno, M. (1992) *Fuzzy Systems Theory and its Applications* (New York: Academic Press).

Thaler, R. (1987) 'The psychology of choice and the assumptions of economics', in A.V. Roth (ed.) *Laboratory Experimentation in Economics* (Cambridge: Cambridge University Press) ch. 4.

Theil, H. (1957) 'A note on certainty equivalence in dynamic planning', *Econometrica*, 25: 346–9.

—— (1964) *Optimal Decision Rules for Government and Industry* (Amsterdam: North Holland).

Theocracharis, R.D. (1961) *Early Developments in Mathematical Economics* (London: Macmillan) ch. 3.

Thomas, L.C. (1984) *Games, Theory and Applications* (New York: Wiley).

Tintner, G. (1941) 'The theory of choice under subjective risk and uncertainty', *Econometrica*, 9: 298–304.

—— (1968) *Methodology of Mathematical Economics and Econometrics* (Chicago: University of Chicago Press).

Tobin, J. (1965) 'The theory of portfolio selection', in (eds) *The Theory of Interest Rates* (London: Macmillan).

Todhunter, I. (1865) *A History of the Mathematical Theory of Probability* (London: Macmillan) ch. 11.

Tukey, J.W. (1971) 'Lags in statistical technology', *Proceedings of 'Statistics '71 Canada'*, 96–104.

—— (1980) 'We need both exploratory and confirmatory', *The American Statistician*, 34 (1): 23–5

Tversky, A. (1969) 'Intransitivity of preferences', *Psychological Review*, 76: 31–48.

Tversky, A. and Kahneman, D. (1971) 'Belief in the law of small numbers', *Psychological Bulletin*, 76: 105–10.

—— (1973) 'Availability: a heuristic for judging frequency and probability', *Cognitive Psychology*, 5: 207–32.

—— (1974) 'Judgment under uncertainty: heuristics and biases', *Science*, 185: 1124–31.

—— (1981) 'The framing of decisions and the psychology of choice', *Science*, 211: 453–8.

—— (1982) 'Judgments of and representativeness', in D. Kahneman, P. Slovic and A. Tversky (eds) *Judgment Under Uncertainty: Heuristics and Biases* (Cambridge: Cambridge University Press).

—— (1986) 'Rational choice and the framing of decisions', in R.M. Hogarth and M.W. Reder (eds) *Rational Choice* (Chicago: University of Chicago Press) 67–94.

Tversky, A., Slovic, P. and Kahneman, D. (1990) 'The causes of preference reversal', *American Economic Review*, 80: 204–17.

Vercelli, A. (1992) 'Causality and economic analysis: a survey', in A. Vercelli and N. Dimitri (eds) *Macroeconomics: A Survey of Research Strategies* (Oxford: Oxford University Press) ch. 14.

von Mises, R. (1957) *Probability, Statistics and Truth* (New York: Wiley) 2nd revised English edition.

von Neumann, J. and Morgenstern, O. (1944) *Theory of Games and Economic Behavior* (Princeton, NJ: Princeton University Press). 2nd and 3rd editions released in 1947 and 1953.

von Winterfeldt, D. and Edwards, W. (1986) *Decision Analysis and Behavioral Research* (Cambridge: Cambridge University Press).

Wakeford, R., *et al.* (1989) 'Childhood leukaemia and nuclear installations', *Journal of the Royal Statistical Society*, A152: 61–86.

Wald, A. (1939) 'Long cycles as a result of repeated integration', *American Mathematical Monthly*, 46: 136.

Walley, P. (1991) *Statistical Reasoning With Imprecise Probabilities* (London: Chapman & Hall).

Wallsten, T.S. (1990) 'The costs and benefits of vague information', in R.M. Hogarth (ed.) *Insight in Decision Making* (Chicago: University of Chicago Press) ch. 2.

Watson, S.R. and Buede, D.M. (1987) *Decision Synthesis* (Cambridge: Cambridge University Press).

Weatherford, R.(1982) *Philosophical Foundations of Probability Theory* (London: Routledge & Kegan Paul).

Weber, M. and Camerer, C.F. (1987) 'Developments in modelling preferences under risk', *OR Spektrum*, 8: 139–51.

Whittle, P. (1963) *Prediction and Regulation by Linear Least-Squares Methods* (London: English Universities Press).

—— (1970) *Probability* (Harmondsworth: Penguin).

Wiener, N. (1949) *Extrapolation, Interpolation and Smoothing of Stationary Time Series* (Cambridge, MA: MIT Press).

Wierzchon, S.T. (1982) 'On fuzzy measure and fuzzy integral', in M.M. Gupta and E. Sanchez (eds) *Fuzzy Information and Decision Processes* (Amsterdam: North-Holland) 79–86.

Wilkinson, G.N. (1977) 'On resolving the controversy in statistical inference', *Journal of the Royal Statistical Society*, B140: 119–71 (including discussion).

Wold, H.O.A. (1953) *Demand Analysis* (New York: Wiley).

—— (1954) *A Study in the Analysis of Stationary Time Series* (Stockholm: Almqvist & Wiksell) 2nd edition.

—— (1964) *Econometric Model Building, Essays on the Causal Chain Approach* (Amsterdam: North Holland).

—— (1991) 'Soft modeling: the basic design and some extensions', in K. Kaul and J.K. Sengupta (eds) *Essays in Honor of Karl A. Fox* (Amsterdam: Elsevier Science) 399–470.

Wold, H.O.A. and P. Faxer (1957) 'On the specification error in regression analysis', *Annals of Mathematical Statistics*, 28: 265–7.

Working, E.J. (1927) 'What do statistical demand curves show?', *Quarterly Journal of Economics*, 41: 212–37.

Yaari, M.E. (1969) 'Some remarks on measures of risk aversion and their uses', *Journal of Economic Theory*, 1: 315–29.

Yager, R.R. (1982) *Fuzzy Set and Possibility Theory* (Oxford: Pergamon Press).

Yager, R.R., Ovchinnikov, S., Tong, R.M. and Nguyen, H.T. (1987) *Fuzzy Sets and Applications: Selected Papers by L.A. Zadeh* (New York: Wiley).

Yule, G.U. (1926) 'Why do we sometimes get nonsense correlations between time-series?', *Journal of the Royal Statistical Society*, 89: 1–64.

Zadeh, L.A. (1962) 'From circuit theory to system theory', *Proceedings of the Institute of Radio Engineers*, 50: 856–65.

—— (1965) 'Fuzzy sets', *Information and Control*, 8: 338–53.

—— (1968) 'Probability measures of fuzzy events', *Journal of Mathematical Analysis and Applications*, 10: 421–7.

—— (1973) 'Outline of a new approach to the analysis of complex systems and decision processes', *IEEE Transanctions on Systems, Man and Cybernetics*, SMC-3: 28–44.

—— (1975a, b) 'The concept of a linguistic variable and approximate reasoning', Parts 1 and 2, *Information Sciences*, 8: 199–249, 301–57.

—— (1976a) 'A fuzzy-algorithmic approach to the definition of complex and imprecise concepts', *International Journal of Man-Machine Studies*, 8: 249–91.

—— (1976b) 'The concept of a linguistic variable and approximate reasoning', Part 3, *Information Sciences*, 9: 149–94.

—— (1978) 'Fuzzy sets as a basis for a theory of possibility', *Fuzzy Sets and Systems*, 1: 3–28.

—— (1984) 'Coping with the imprecision of the real world: an interview with Lofti A. Zadeh', *Communications of ACM*, 27: 309–11. Reprinted in Yager et al. (1987) 9–28.

—— (1986) 'Is probability theory sufficient for dealing with uncertainty in AI: a negative view', in L.N. Kanal and J.F. Lemmer (eds) *Uncertainty in Artificial Intelligence* (Amsterdam: North-Holland) 103–16.

Zeckhauser, R. (1986) 'Comments. Behavioral versus rational economics: what you see is what you conquer', *Journal of Business*, 59 (4, Part 2): S435–49.

Zellner, A. (1971) *An Introduction to Bayesian Inference in Econometrics* (New York: Wiley).

—— (1979) 'Statistical analysis of econometric models', *Journal of the American Statistical Association*, 74: 628–651, including discussion.

—— (1988a) 'Bayesian analysis in econometrics', *Journal of Econometrics*, 37 (1): 27–50.

—— (1988b) 'Bayesian analysis in econometrics', in W. Griffiths, H. Lutkepohl and M.E. Bock (eds) *Readings in Econometric Theory and Practice*, ch. 4.

Zimmermann, H.-J. (1991) *Fuzzy Set Theory – and Its Applications* (Boston: Kluwer Academic Press) 2nd Edition.

NAME INDEX

192

196

SUBJECT INDEX

198

SUBJECT INDEX

DATE DUE
